UNITED STATES
MARTIAL PISTOLS
AND REVOLVERS

UNITED STATES MARTIAL PISTOLS
AND
REVOLVERS

A Reference and History

BY

ARCADI GLUCKMAN

Skyhorse Publishing

Skyhorse Publishing books may be purchased in bulk at special
discounts for sales promotion, corporate gifts, fund-raising, or
educational purposes. Special editions can also be created to
specifications. For details, contact the Special Sales Department,
Skyhorse Publishing, 307 West 36th Street, 11th Floor, New York,
NY 10018 or info@skyhorsepublishing.com.

Skyhorse® and Skyhorse Publishing® are registered trademarks of
Skyhorse Publishing, Inc.®, a Delaware corporation.

Visit our website at www.skyhorsepublishing.com.

10 9 8 7 6 5 4 3 2 1

Library of Congress Cataloging-in-Publication Data is available on file.

Cover design by Richard Rossiter

Print ISBN: 978-1-62914-440-5
Ebook ISBN: 978-1-63220-167-6

Printed in the United States of America

EDITOR'S NOTE

ARCADI GLUCKMAN'S

UNITED STATES
MARTIAL PISTOLS AND REVOLVERS

by Dr. Jim Casada

Born in 1896, Arcadi Gluckman was a career military man who rose to the rank of colonel in the United States Army. Biographical information on him is sparse. He was retired by 1964, when Harold L. Peterson's *Encyclopedia of Firearms* appeared. Peterson notes as much in his coverage of contributors. It seems likely that there would have been obituary mentions in firearms-related magazines when Gluckman died, but if so, I have been unable to locate them. Most of his considerable literary work focused on military handguns, the history of gun making, and the evolution of American long guns. He was widely recognized as a leading authority in the field.

In addition to the work reprinted here, Gluckman wrote or coauthored a number of other books. These include *United States Muskets, Rifles and Carbines* (later revised and retitled *Identifying Old U.S. Muskets, Rifles & Carbines*); *American Gun Makers* (with Leroy De Forest Satterlee — an important reference work in dictionary form, with commentaries on individual gun makers); *Catalogue of United States Martial Short Arms* (with assistance from George H. Smoots); and a special limited edition, slipcased book, *The Collecting of Guns*. Gluckman was also a frequent contributor to gun-related periodicals, wrote entries for Peterson's *Encyclopedia of Firearms,* and served as a consultant to collectors.

United States Martial Pistols and Revolvers has an interesting and complex publishing history. It was first published in 1939 by the Otto Ulbrich Company of Buffalo, New York. It was bound in strawberry-colored cloth. Some sources say there was no dust jacket, while others list a cover. I have never seen the book with a dust jacket. Copies of the first edition have become quite difficult to find. In addition to this trade edition, the work was also offered in a limited, signed edition of one hundred copies specially bound in three-quarter morocco over coarse brown cloth. The lettering on the spine in this version is gilt, as is the top edge. The deluxe edition is extremely rare and highly prized by collectors. The above-mentioned *Catalogue of United States Martial Short Arms,* which evidently came in two forms (spiral binding and case binding with a dust jacket), appeared in the same year under the same imprint, and although it is a distinct and separate work, clearly it was intended to serve as a companion to *United States Martial Pistols and Revolvers.*

Thanks no doubt in part to the entry of the United States into World War II not too long after its publication — a time of national upsurge in interest in all things military — the book enjoyed considerable early success. This was sufficient to engender a 1944 reprint, in the same format and from the same publisher. A dozen years later, in 1956, there were yet more reprints, again in the same format, from the Stackpole Company and Bonanza Books (both paperbound and hardback). Bonanza reprinted the book again in 1961, while by that date Stackpole was in its fourth printing. A generation after the book's initial appearance, in 1974, Crown Publishers brought out yet another hardback reprint. Then in 2007 and again in 2010, Kessinger reprinted the work in softbound form, and in 2011 offered it in hardback.

The work is a vital reference source from a number of perspectives. Its listing of single-shot pistols, revolvers, and semi-automatic pistols, covering the period 1799–1917, is invaluable for collectors. Most appealing to the average reader are its brief histories of well over a hundred handguns. As a landmark in the field, it is still a useful reference and a work of enduring interest.

Jim Casada

ROCK HILL, SOUTH CAROLINA

CONTENTS

Part I

UNITED STATES MARTIAL SINGLE SHOT PISTOLS

Chapter I

UNITED STATES MARTIAL FLINTLOCK PISTOLS

THE DEVELOPMENT OF THE FLINTLOCK PISTOL— EARLY IGNITION SYSTEMS — FUNCTIONING OF THE FLINTLOCK PISTOL — VARIATIONS FROM MODELS — CALIBERS AND GAUGES — COMMITTEE OF SAFETY AND CONGRESSIONAL PIS-TOLS — THE FIRST U. S. MARTIAL PISTOLS — AN OUT-LINE OF SIMEON NORTH HISTORY — MODEL 1799 NORTH & CHENEY — MODEL 1806 HARPERS FERRY — MODELS 1808, 1810 NORTH BERLIN — MODELS 1813, 1816 NORTH MIDLN — MODEL 1816 NORTH VARIATION — MODEL 1818 SPRING-FIELD — MODELS 1819, 1826 NORTH — MODEL 1826 EVANS — MODELS 1836 R. JOHNSON, A. WATERS, A. H. WATERS & CO.

Chapter 2

UNITED STATES MARTIAL PERCUSSION PISTOLS

THE DEVELOPMENT OF THE PERCUSSION SYSTEM — INCENDIARY MIXTURES — GUNPOWDER — FIRST CARTRIDGES — FULMINATES — FORSYTH LOCK — U. S. CONVERSIONS — MAYNARD TAPE PRIMER — U. S. PERCUSSION PISTOLS — MODELS 1842 H. ASTON, H. ASTON & CO., I. N. JOHNSON, PALMETTO ARMORY — MODELS 1843 N. P. AMES, DERINGER — MODEL 1855 SPRINGFIELD PISTOL-CARBINE — HARPERS FERRY PISTOL-CARBINE.

Chapter 3

UNITED STATES MARTIAL SINGLE SHOT

CARTRIDGE PISTOLS

THE DEVELOPMENT OF THE METALLIC CART-RIDGE — MODELS 1866, 1867 REMINGTON — MODEL 1869 SPRINGFIELD — MODEL 1871 REMINGTON.

Chapter 4

UNITED STATES SECONDARY MARTIAL PISTOLS

NOTES ON SECONDARY MARTIAL PISTOLS — AN-STAT—BIERLY & CO.—C. BIRD & CO.—BOOTH—CAL-DERWOOD MODEL 1808 TYPE — T. P. CHERINGTON — COUTTY — H. DERINGER MODEL 1808 TYPE — H. DER-INGER MODEL 1826 TYPE — JOHN DERR — DREPPERD PERCUSSION — ELGIN PERCUSSION CUTLASS-PISTOL (C. B. ALLEN MAKE) — ELGIN PERCUSSION CUTLASS-PISTOL (MER-RILL, MOSMAN & BLAIR MAKE) — EVANS FRENCH MODEL 1805 TYPE — T. FRENCH MODEL 1808 TYPE — T. GRUBB — I. GUEST MODEL 1808 TYPE — HALL BREECH-LOADING FLINT-LOCK (BRONZE BARREL AND BREECH) — HALL BREECH-LOADING FLINTLOCK (IRON BARREL AND BREECH) — J. HENRY (PHILA) MODEL 1808 TYPE — J. HENRY — J. J. HENRY (BOULTON) — J. J. HENRY (BOULTON) MODEL 1826 TYPE — C. KLINE — KUNTZ — LINDSAY TWO SHOT PERCUSSION — MARSTON BREECH-LOADING PERCUSSION — McK BROTHERS (BALTIMORE) PERCUSSION — MEACHAM & POND — MILES MODEL 1808 TYPE — MILES — B. MILLS — P. & D. MOLL — I. PERKIN — PERRY BREECH-LOADING PERCUSSION — POND — RICHMOND, VIRGINIA — ROGERS & BROTHERS — JOHN RUPP — SHARPS BREECH-LOADING PER-CUSSION — SHULER MODEL 1808 TYPE — SWEITZER MODEL 1808 TYPE — VIRGINIA MANUFACTORY — J. WALSH — A. H. WATERS & CO. PERCUSSION.

Part II

UNITED STATES MARTIAL REVOLVERS
AND
AUTOMATIC PISTOLS

Chapter 1

MARTIAL PERCUSSION REVOLVERS

PERCUSSION REVOLVERS — An outline of Colt history—Combustible cartridges—Instructions for loading percussion revolvers. ADAMS —ALLEN & WHEELOCK—ALSOP — BEALS — BUTTERFIELD — COLT Models 1839, 1847, 1848, 1851, 1855, 1860, 1861, 1862 — COOPER — FREE-MAN—JOSLYN — LEAVITT — MANHATTAN—METRO-POLITAN—PETTINGILL—PLANT—REMINGTON Model 1861, New Model — REMINGTON-RIDER — ROGERS & SPENCER — SAVAGE-NORTH — SAVAGE — STARR — UNION—WALCH—WARNER—WESSON & LEAVITT—WHITNEY.

Chapter 2

MARTIAL CARTRIDGE REVOLVERS

BACON — COLT Models 1872, 1878, 1892-94-96, 1901-03 Army, 1889-95 Navy, 1907, 1909, 1917 — FOREHAND & WADSWORTH — HOPKINS & ALLEN — MERWIN & HULBERT — PLANT — POND — PRESCOTT — REM-INGTON Model 1874 — SMITH & WESSON Models No. 2, 1869, 1875, 1881, 1899, 1917.

Chapter 3

AUTOMATIC PISTOLS

AUTOLOADING PISTOLS — COLT Military Model 1902, Model 1911—*GRANT-HAMMOND—SAVAGE* Model 1905.

FOREWORD

This volume on the subject of United States martial pistols and revolvers is presented in an attempt to fill a long felt need of the collector and "gun crank" for a work containing within single covers, the development and description of our martial short arms from the flintlock pistol of the early period as a nation, to the latest issue of the World War. Though most of our martial pistols and revolvers have been described individually in a number of separate articles, monographs, pamphlets and volumes, the data appears in so great a variety of publications scattered over a number of years; — some in limited editions and most of them long out of print; — that the search for information to aid in identification, classification and study of the available historical background has become a task of considerable magnitude and beyond the resources of the average lover of our military firearms. It is to this service that this book is dedicated.

In addition to the pistols made in government armories, or under government contract for use in the military or naval service, or for issue to the militia of the several states, pistols of military type, caliber and size, made for sale to individual officers, privateers and militia by private makers of their own initiative, are treated in a separate chapter on U. S. Secondary pistols.

The details and measurements of the arms described were based on the study of specimens in original and fine condition in the author's collection, or made available to him through the courtesy of other collectors. Any comparison of flintlock arms for purposes of identification must allow for the fact that many of the arms of that period were made by hand, sometimes, and especially in U. S. Secondary pistols, entirely by the individual craftsman, and in the same model minor variations, such as barrel length, marking, and often details of construction, were a common occurrence. Another source of differences lay in local repairs and replacements of worn out, damaged, or lost parts. Again with the introduction of new models, interchangeable old model parts in stock were used until exhausted, creating models with a "bar sinister" on their family escutcheon.

The author desires to express his grateful appreciation to collectors, authors, publishers and dealers who contributed material and data for this work, and whose friendly interest, and valuable assistance made this volume possible.

Special acknowledgment is made to Mr. L. D. Satterlee for generous contribution of historical notes; — to Messrs. Francis W. Breuil, S. Harold Croft, J. C. Harvey, Charles T. Haven, Thomas T. Hoopes, L. D. Ingalls, William G. Renwick, Sam E. Smith, Dr. Thomas B. Snyder and Yale University, for invaluable assistance and access to their collections; — and to Mr. Thomas G. Samworth for helpful suggestions and advice.

Corrections and additional data sent in care of the publisher will be appreciated and incorporated in future editions.

ARCADI GLUCKMAN,
Major Infantry, U. S. Army.

UNITED STATES MARTIAL SINGLE SHOT PISTOLS

Chapter I.

UNITED STATES MARTIAL FLINTLOCK PISTOLS

THE DEVELOPMENT OF THE FLINTLOCK PISTOL— DEMIHAGUE — SERPENTINE — MATCHLOCK — WHEEL LOCK — SNAPHANCE — FLINTLOCK — MIGUELET — FIRELOCK — THE FUNCTIONING OF THE FLINTLOCK PISTOL — VARIATIONS FROM MODELS — CALIBERS AND GAUGES — COMMITTEE OF SAFETY AND CONGRESSIONAL PISTOLS — THE FIRST U. S. MARTIAL PISTOLS —

RAPPAHANNOCK FORGE —

MODEL 1799 NORTH & CHENEY, BERLIN —

MODEL 1806 HARPERS FERRY—

MODEL 1808 NORTH, — BERLIN

MODEL 1810 NORTH, — BERLIN

MODEL 1813 NORTH, — MIDLn CON.

MODEL 1816 NORTH, MIDLn. CON.

MODEL 1816 NORTH, MIDLtn., CONN.

MODEL 1818 SPRINGFIELD —

MODEL 1819 NORTH —

MODEL 1819 NORTH —

MODEL 1826 NORTH —

MODEL 1826 EVANS —

MODEL 1836 R. JOHNSON —

MODEL 1836 A. WATERS —

MODEL 1836 A. H. WATERS & CO. —

Part I.

UNITED STATES MARTIAL SINGLE SHOT PISTOLS.

Chapter I.

UNITED STATES MARTIAL FLINTLOCK PISTOLS

The DEVELOPMENT of the FLINTLOCK PISTOL

It is generally reputed that the pistol as a hand arm was invented by Caminelleo Vitalli at Pistoia, Italy, in 1540, the weapon being named for the city of its origin. The word "pistol" is mentioned by Louis de Gaya, cele-brated French arms expert of the 17th Century, as early as 1678 in his treatise "Arms and Engines of War". However hand guns were known and used in warfare long prior to 1540: — a crude bronze tube with a touch-hole at the top, marked 1322, was unearthed at Mantua, in Italy, another in the Castle of Monte Vermini, was dated 1346; while a third stamped 1364, was found in Perugia.

The DEMIHAGUE

It was not until the end of the 14th Century however, that the harquebus came into general use as a weapon of warfare. Harquebus is of German origin meaning "a gun with a hook", the hook's function being to reduce the recoil and steady the piece while fired, by resting against a rampart or a portable rest. The fore-runner of the pistol was the demihague, about half the size and weight of the harquebus, a short and crude weapon used by the cavalry-man. It had a ring at the end of the stock with a cord

THE DEMIHAGUE
CIRCA 1450
FROM CODEX LATINUS 7239 LIBRARY OF PARIS

running through for suspension from the saddle bow or around the trooper's neck. The demihague was steadied on a forked rest fastened to the saddle bow, and was dis-charged by igniting the powder at the touch-hole by means of a wick-like "slow match" made of cotton or hemp, and boiled in a strong solution of saltpetre. From the touch-hole the fire flashed to the main charge in the closed end of the crude tube, discharging the piece with a roar, fire and smoke, that were just as effective in destroying enemy morale as the effect of the projectile itself.

THE SERPENTINE

The difficulty of holding the gun with one hand while applying the "match" with the other, led to the invention of the serpentine, some time between 1460 and 1470. The serpentine was an "S" shaped iron rod pivoted

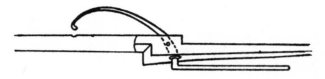

EARLY SERPENTINE LOCK

through, or on the side of the stock, which with a movement of a finger brought the opposite end holding the glowing match, into the touch-hole on the top of the breech.

THE MATCHLOCK

The constant danger from a flash-back atop the gun led to the early invention of the matchlock, in which a split-tip, crooked lever holding the match was moved by means of a pivoted lever or trigger, against the action of a spring, to ignite the priming powder in a pan on the *side* of the barrel. In time the lock was further refined by the addition of a spring, which upon pressure on the trigger, accelerated the fall of match-holder to the flash-pan. A fence in rear of the pan provided additional safety.

Though the matchlock was an improvement on the

simple serpentine lock, its user was still inconvenienced by the necessity of carrying about him several feet of slow

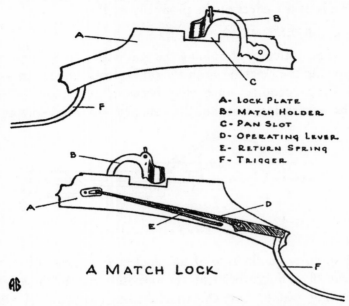

A - LOCK PLATE
B - MATCH HOLDER
C - PAN SLOT
D - OPERATING LEVER
E - RETURN SPRING
F - TRIGGER

A MATCH LOCK

burning match; was exposed to the danger of burns and accidental discharge of the piece, as well as possible ignition of the loose powder carried as a part of the equipment. However in spite of these dangers and the difficulty of keeping the match and the flash-pan dry in wet weather, the mechanism of the matchlock was so simple to make and so rugged in use, that although eventually supplanted by the wheel lock (believed to have been invented in Nuremberg, Germany in 1517), the matchlock remained in use, concurrently with the wheel lock, for another hundred and fifty years as the principal infantry firearm mechanism. Matchlocks were used as early as 1471, when Edward IV used a force of matchlock armed sharpshooters

when landing in England to regain his throne. Their last
known military use was in 1680 during the Monmouth
Rebellion in the West of England. Long, slender matchlocks are still used by the native hunters of China, Thibet
and India.

The WHEEL LOCK

The wheel lock consisted of a grooved and notched
steel disc, the serrated edge protruding into a covered
flash-pan. The wheel was pivoted on an arbor or spindle,
whose protruding square end could be wound with a
key or spanner against the action of a powerful spring
attached to the spindle by a short chain and held under
tension by a sear. A short lever (dog-head) holding a
piece of iron pyrites (metallic sulphide) in its adjustable
jaws, was made to press on the sliding flash-pan cover by
means of a spring. When the trigger was pulled, the
released spring, acting against the chain wound around the
spindle, caused the wheel to revolve rapidly, the pan cover
slid back mechanically, and the contact of the iron pyrites
against the whirling, serrated steel threw sufficient sparks
into the pan to ignite the priming powder and discharge
the piece.

The improvement of the lock enabling the piece to be
carried in a holster primed and ready for firing, led to
further refinement of the arm. The first types of pistols
developed in the early 16th Century, with their short
heavy barrels and thick angular butts, gave way by 1580
to slender barrels, lengthened butts, and locks and stocks
embellished with engraving and inlay.

Though the wheel lock was a vast improvement on the

matchlock, it was complicated in operation, difficult to maintain in serviceable and reliable working condition, and was slow in use, requiring winding up with a key for every shot. The cost of this elaborate mechanism precluded its general military use. Though the matchlock was simple and cheap, an ignition system which required the use of a lighted "match" at all times, and whose serviceability depended on weather conditions, was also extremely unsatisfactory. A simpler, cheaper and more reliable ignition system was needed and was provided in the flintlock, through gradual development of the ignition systems.

THE SNAPHANCE

The snaphance, or snaphaunce, the fore-runner of the flintlock, was invented towards the latter part of the Sixteenth Century. In principle it placed a flint in the jaws of a hammer, which when released by pressure on the trigger, struck a downward, glancing blow against the battery (frizzen), a hinged piece of steel held in position over the pan. The resulting sparks ignited the priming charge in the pan and discharged the piece. The mechanism and mainspring of the snaphance were generally placed inside the lock-plate, the frizzen spring was located outside. To protect the priming powder the pan was equipped with a sliding pan cover, manually closed after loading, but mechanically opened before firing, by the fall of the hammer.

Authorities differ as to the country of its origin. Some claim that is was a Dutch invention, from the words "snaap-haans", meaning "chicken thieves" and such gentry, who found the wheel lock too expensive, and the matchlock too betraying in their nocturnal operations

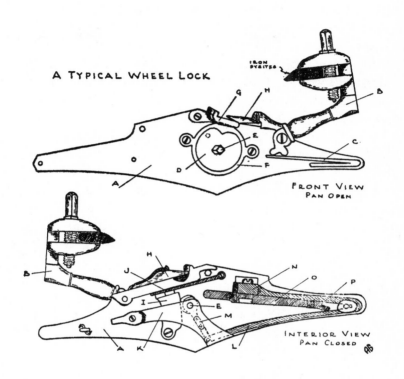

A TYPICAL WHEEL LOCK

FRONT VIEW
PAN OPEN

INTERIOR VIEW
PAN CLOSED

A—Lock plate

B—Doghead (hammer)

C—Doghead spring

D—Wheel

E—Spindle or wheel arbor

F—Wheel housing ring

G—Flash pan

H—Pan cover

I—Pan cover shaft

J—Pan cover friction spring

K—Wheel arbor bridle

L—Mainspring

M—Chain

N—Sear bridle

O—Primary

P—Secondary Sear

because of light and smell, and so of necessity, were responsible for its invention and development. Others lay a similar claim to its origin in the Low Countries, (Holland and Flanders) from the German "schnapp hahn", or pecking fowl, descriptive of the action of a hen pecking, from the fall of the flint holding hammer. Still others, pointing to the fact that it is also known as the "Spanish Lock", allege that "snaphance" is but the Anglicized version of "snapphahn", meaning snap-lock, and was brought to England by the English forces campaigning in the Low Countries, where the lock had been introduced by Spanish troops in the wars against the Dutch. They concede however, that the lock is the product of the inventive brain of thieves and brigands who found the earlier ignition systems unsuitable to nefarious night operations and ambuscades.

Probabilities are that the snaphance and its variations were the development of more countries than one, in effort to find a mechanical application of ignition by means of striking a flint against steel, at that time the universal method of making fire.

THE FLINTLOCK

By the middle of the Seventeenth Century the snaphance was improved into what is now known as the flintlock, by combining the pan cover and battery into one piece, the frizzen, held in place over the pan against the action of a spring, and pivoted to tilt out of the way when struck by the flint. Another important improvement was the additional safety feature provided by the invention of the half-cock.

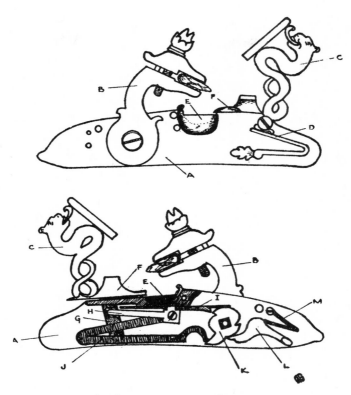

A SNAPHANCE LOCK
LATE FORM

A—Lock plate

B—Cock or hammer

C—Battery or frizzen

D—Frizzen spring

E—Pan

F—Pan cover (sliding)

G—Pan cover shaft

H—Pan cover friction spring

I—Pan cover opening bar

J—Main spring

K—Tumbler

L—Sear

M—Sear Spring

The MIGUELET

A Spanish variation of the early flintlock, known as the miguelet, differed chiefly in the external details of the mechanism, such as the main spring placed outside the lock-plate. The designation "miguelet lock" or "miguelet", as it is sometimes called, is said to have come from the Spanish or Portuguese brigands known as "miquelitos", who are reputed to have carried such arms.

The FIRELOCK

Early Colonial and English records often mention the firelock which was first used to distinguish a weapon discharged by igniting the priming in the flash-pan by sparks mechanically, from the matchlock which was ignited by a slow-burning match. Though primarily applied to wheel locks, with the development of the snaphance and the flintlock, the term firelock was also applied to these ignition systems.

The FUNCTIONING of FLINTLOCK ARMS

The functioning of the smoothbore flintlock pistol may be described as follows:—

To load:—The pistol was loaded by the use of a prepared cylindrical paper cartridge containing a suitable load of black powder and a soft-lead, spherical ball, the ends of the cylinder fastened by twisting or paste. The hammer was placed at half-cock, the pan opened by throwing the frizzen forward, the end of the cartridge containing the powder was bitten off, a pinch of powder squeezed from the open end into the priming-pan, and the pan closed by lowering the frizzen. The remainder of the powder was

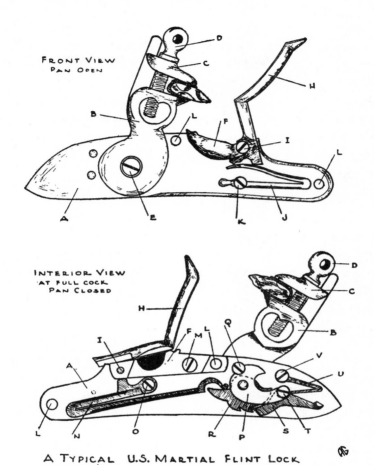

FRONT VIEW
PAN OPEN

INTERIOR VIEW
AT FULL COCK
PAN CLOSED

A TYPICAL U.S. MARTIAL FLINT LOCK

A—Lock plate
B—Hammer
C—Cap
D—Hammer screw
E—Tumbler screw
F—Pan
G—Pan screw
H—Frizzen
I—Frizzen screw
J—Frizzen spring
K—Frizzen spring screw

L—Side screw holes
M—Pan screw
N—Main spring
O—Main spring screw
P—Bridle
Q—Bridle screw
R—Tumbler
S—Sear
T—Sear screw
U—Sear spring
V—Sear spring screw

then poured into the barrel, the round ball rammed on the powder, and the remaining paper crumpled and rammed on top of the bullet to keep the loose ball from falling out. The loading of the rifled pistol was similar, except that a greased leather or linen patch was used to facilitate the ramming home of the bullet and prevent its distortion by forcing against the resistance of the rifling. Of course, loose powder carried in a powder horn could also be used.

To fire:—The hammer was cocked. When the trigger was pressed, the sear released the tumbler which revolved forward by the action of the main spring. The hammer fell, and a flint held in the hammer jaws struck at an angle against the steel frizzen and tilted it forward, throwing sparks into the pan which was uncovered by the forward tilt of the frizzen. The sparks ignited the priming charge, and the fire flashed through the touch-hole to the main charge causing ignition and discharge.

The flints used in our service were generally obtained in England or France. The best flints were translucent, with a smooth surface, of uniform tint of light yellow or brown color. In the U. S. service the flint was set in the hammer with the bevel uppermost clamped in a folded strip of lead or leather, though in some locks it was necessary to set the flint flat side up to obtain the best results in throwing sparks and tilting the frizzen. A good flint could be used about sixty times. Flints were issued to troops in the proportion of one flint to every twenty rounds.

It is to be borne in mind that in spite of the vast improvement in the ignition mechanism of the flintlock pistols of the end of the Seventeeth Century over their

ancestors of Pistoria of 1540, they were still weapons useful only in hand-to-hand encounter. The relatively long
time that elapsed between the pressing of the trigger and
the discharge of the ball, precluded the possibility of
steady aim; — and the inaccuracy of the loosely loaded
ball dispensed with the necessity of sights for accurate aim.
Many of the martial pistols of the period, such as Models
1799, 1808, 1810 and 1813 were not equipped with
sights. When the sights were provided on later models,
the pistols were generally fitted with one sizeable sight in
front only, but its function was more to assist the eye in
picking up the muzzle quickly than to align the barrel.
While successful experiments had been made with multi-
barreled or multi-fired small arms; not only in flintlocks
but even in wheel locks, these were rare, delicate and expensive and unsuitable for military use. Therefore in order
to provide for a second shot in case of misfire or miss, martial pistols were usually carried in pairs; — by the army
in leather holsters on each side of the saddle pommel; —
by the navy in braces, thrust in sashes or belts.

As mentioned in the preface, the details, measurements and illustrations to follow are based on the study of
the best available specimens in original and fine condition,
from the author's collection, or made available to him
through the courtesy and co-operation of other collectors.
The ardent gun-lover, and collector of our historical pistols and revolvers, using this volume in attempt to identify
some lucky find or treasured heirloom, will often find some
minor variation in the length of barrel, the marking, or a
detail of construction. Any comparison with the case
specimens described, must allow for the fact that many of

the early arms, especially in the Secondary U. S. pistols class, were made entirely by hand, often entirely by one individual, with resulting variations even in the same model. Worn out, broken or lost parts were usually replaced by local gunsmiths with parts similar and service-able, but not original. Another source of differences lay in the continued use of suitable old model parts with neces-sary modifications, upon introduction of new models, until the old stock of parts was exhausted. Other deviations from the accepted model were caused by changes, modifi-cations and improvements demanded by the War Depart-ment prior to the completion and delivery of all the pistols made under contract, which usually stretched over a period of a few years, during which the original models of the early deliveries were tested under service conditions and found wanting.

The dating of the arms described in this volume is based on the usual procedure of accepting the date of the approval of the pattern by the War Department, or the contract date in contract arms, as the date of the model, insofar as such information has been available at the time of writing.

It might also be pertinent to mention at this point that the calibers of the early martial arms were based not on the diameter of the bullet but on the weight of the spheri-cal balls to the pound. The standard musket projectile was an ounce ball of 16 gauge (16 to the pound) and meas-uring approximately .675 inches in diameter. The other standard caliber was that of the rifle bullet, half ounce in weight, 32 gauge (32 to the pound) measuring about .530 inches. This caliber was also used in pistols.

In theory, the arms using the above calibers, should have had a bore diameter of .69 and .54 respéctively, to allow for clearance in loading, regardless of type of arm. But under conditions of hand manufacture a considerable variance was to be expected, and smiliar models may vary as much as .05 of an inch from the standard.

COMMITTEE OF SAFETY AND CONTINENTAL PISTOLS

In covering the development of the United States martial pistols and revolvers it seems desirable from the historical and chronological standpoint to mention the pistols of our early period as a nation, from the revolutionary era to the last decade of the Eighteenth Century, when the Springfield Armory, the first government arms factory, was established in 1795. During this period contracts were also awarded by the United States Government for the manufacture of arms for the army, the navy and the militia of the several states, which latter were largely supplied by the Federal Government.

The martial pistols used by the Continental troops during the Revolution were limited in number, and were largely of British and French origin, mostly the personal property or equipment of the officers of the Colonial militia. The British pistols had been imported from England during colonial days or issued for use in the French and Indian Wars; the French pistols were those captured during these campaigns, and augmented by the few hundred presented by Lafayette and acquired by purchase.

Shortly before the Revolution, Great Britain, sensing the rebellious attitude of the North American Colonies prohibited the export of arms to the disaffected provinces. Notwithstanding this arms embargo, but little was done

by the colonial authorities to encourage domestic fabrication on a large scale, until the hostilities actually began on
the Lexington Green in 1775. With the outbreak of the
Revolution, provincial authorities hastented to encourage
the manufacture of arms and munitions with offers of
bounties and exemptions. Committees of Safety were
formed in each colonial division, charged with the arming
and equipping their martial levies. These Committees are
known to have purchased such arms as were available in
the hands of the merchants and gunsmiths, and to have
contracted for more with the latter. While there existed
a real need for pistols for the use of officers, mounted mes-
sengers, and for marine use; in the acute need for arms,
the preference for manufacture, contract and purchase of
small-arms was given to the procurement of muskets and
rifles; the pistols of those days being weapons of but limited use. While doubtless pistols were made and supplied
to the Committees by colonial gunsmiths, no records are
available as yet, nor has any pistol been found of the type
of the French or British martial pistols of the period,
which through marking and date would permit positive
identification of the arm as a Committee of Safety pistol.

The nearest approach to the pistols in this classification is the arm made by the Rappahannock Forge which
was established by an act of the Virginia Assembly in June
1775, near Fredericksburg, Virginia, and manufactured
small-arms from the date of its founding until May 30,
1781, when it was dismantled to prevent capture and
destruction by a British raiding party under command of
Lieut. Colonel Banastre Tarleton.

Doubtless the earliest arms which could be classified

with certainty as U. S. martial pistols would be the weapons furnished to officers and to seamen of the Continental forces, and issued to mounted organizations of the army after the establishment of the Republic. It is known that the Continental Congress purchased and contracted for arms to equip the troops and that pistols made by American gunsmiths were included in these purchases and contracts; — but, the remarks pertinent to the lack of positive identification of the Committee of Safety pistols are equally applicable to the Continental pistols. None have been found to date which can be positively identified as such.

The clue available at this time as to these early U. S. pistols is in the inventory of public arms in U. S. storehouses, dated December 14, 1793, which shows 805 serviceable pistols in storage in the arsenals named: —

Arsenal	Pistols
Springfield, Mass.	495
West Point, N. Y.	59
Philadelphia, Pa.	251

In addition to the above, 420 unserviceable pistols are shown in the return, of which 407 were stored at Philadelphia. Unfortunately the inventory does not distinguish these arms as to make and type.*

RAPPAHANNOCK FORGE FLINTLOCK PISTOL

Caliber .69, taking an ounce spherical ball. Nine inch round, smoothbore barrel. Total length fifteen-and-one-eighth inches. Iron mountings. Brass trigger guard. The barrel is pin fastened to the full length stock and is with-

---

* American State Papers, Class V, Military Affairs, Paper 10.

out sights. The butts is bulbous, hexagonal shaped. Double-necked, flat, bevelled hammer. The pan is iron, with flat bevelled edges and has a fence to the rear. Hickory ramrod.

In the specimen described the flat, bevelled-edge lockplate is marked "RAPa FORGE" and "CP". The barrel is unmarked. The stock is marked "GW" on the left side.

The letters "CP" on this arm indicate that the pistol at one time was the property of the Commonwealth of Pennsylvania, whose arms were so marked under Article 1, Section 1, Act of March 28, 1797. (See Abridgement of the Laws of Pennsylvania for Year 1700 to 2nd Day of April 1811, Philadelphia 1811.)

Rappahannock Forge was the alternate name of the Hunter Iron Works established by James Hunter at Falmouth, Va., some time before the Revolution. At the outbreak of the war the works were considerably enlarged; a manufactory of small-arms was added, slitting mills were constructed, and anchors and war material manufactured. The manufactory probably had a contract to make muskets and quite likely made pistols for mounted troops, such as Baylor's, Moylan's and Light Horse Harry Lee's Dragoons.

The factory was dismantled about May 30, 1781, on the approach of a British raiding party under Tarleton. Later the workmen were recalled and the works enlarged, but failing to receive financial support from the State of Virginia, James Hunter dismissed the remaining workmen and closed the factory December 1, 1781.

It is believed that in 1777-78, pistols were purchased in Prussia for the Hunter Iron Works by William Lee, State Agent of Virginia. It is quite possible that some of these Prussian pistols were later marked with the Rappahannock Forge markings.

THE FIRST U. S. MARTIAL PISTOLS

In the years following the War of Revolution, the domestic and foreign conditions confronting the young Republic were unsettled and troublesome. At home the westward march of the young and lusty nation provoked bloody conflicts with the Indian tribes of the border states. Abroad there was serious friction with France even extending to naval engagements between French and American vessels. Our young navy had tested its mettle in the Mediterranean against the Barbary States. American rights at sea were disregarded by both England and France who were at war. Both belligerents had issued orders forbidding neutral trade with its opponent under threat of confiscation. Relations with England were far from friendly due to search of American vessels, and impressment of seamen for service on British men-of-war. So threatening were the conditions, and so serious the shortage of arms, for which there had been practically no replacement since the Revolution, that as early as in 1794 Congress prohibited the export of arms and munitions, and welcomed the import of foreign arms, duty free.

Though a national armory had been established at Springfield, Massachusetts, in 1794, which from a small output of but 245 muskets in the first year of its existence, rapidly increased its production facilities until it reached a year's total of 4595 muskets in 1799;—the fabrication of arms remained inadequate to supply the needs of the army, the navy and the militia. The urgent need for long arms precluded the manufacture of pistols at the newly built armory for some years to come. To remedy this situation the Congress, in 1798, appropriated $800,000, a very considerable sum in those days, for the purchase of arms and ammunition to augment the output of the Springfield Armory. Contracts were made with private firms, gunsmiths and armorers for the fabrication of military equipment, muskets, swords and pistols for the government. The pistol contract for five hundred pistols was awarded on March 9, 1799, to Simeon North of Berlin, Connecticut; who thereby became the first official pistol maker to the United States.

Simeon North, a descendant of an old New England family, was born at Berlin, Connecticut, July 13, 1765, and was sixteen years old at the time of Cornwallis' surrender at Yorktown. Family history states that young North began life as a farmer. In 1795 he purchased a water power saw mill adjoining his farm, in which he started the manufacture of scythes, at that time an indispensable item of farming equipment. It was probably this training as a metal worker and machinist, and a natural mechanical bent, that prompted him to secure a government contract on March 9, 1799, for the manufacture of horse pistols. Possibilities are that he learned the rudiments of pistol manufacture from a gunsmith neighbor, Elias Beckley, whose gun shop was but a mile away from North's birthplace in Berlin.

The first North pistols (Model 1799) were so satis-factory, and the need was so great, that even prior to the completion and delivery of the first 500 pistols, North was awarded another contract for 1,500 pistols, to be completed by Feb. 6, 1802.

After completion of these contracts North resumed the manufacture of farm implements until June 30, 1808, when he obtained a contract for 2,000 navy boarding pis-tols to be made according to Navy Department specifica-tions and patterns, but with North improvements of an iron back-strap, a brass pan instead of the easily corroded iron pan of the pattern pistol, as well as some advantage-ous minor alterations. To fulfill this contract North enlarged his factory and applied his mechanical ability and inventive genius to the development of labor-saving machinery and the modern principle of standardization of parts, by assigning the production of identical parts to individual workmen, until a large number were finished. This method, far ahead of the times of individual crafts-manship, resulted in more uniform and better product as well as saving in time and labor, though it compelled North to purchase all materials and manufacture all components before finishing a single pistol of any batch to be delivered.

The pistols of this contract (Model 1808) having been "much approved", on Dec. 4, 1810, North was given a contract by the Navy Department for an addi-tional thousand.

In the meantime, in 1808, the Congress passed an act for the arming and equipping the whole body of the militia of the several states by the Federal Government, and though an additional government armory had been established at Harpers Ferry, Virginia, in 1796, large contracts for arms were placed with numerous private arms manufacturers and armorers.

Simeon North's first contract under the Act of 1808 was made in 1810 with Tench Coxe, Purveyor of Public Supplies, and was for manufacture of horse pistols for the army (Model 1810). Unfortunately no data is available as to the production under this contract, as the contract papers have not been located in the government files. Probabilities are however, that the original contract was for not less than 5,000 pistols, as records show that in 1813 North still owed 2,400 pistols on this account.

The earliest North pistols made from 1799 to 1802, and patterned after the French Model 1777, are marked with Cheney's name as well as North's. In spite of this marking the North family history denies any partnership in Simeon North's establishment, though a working agree- ment is acknowledged to have existed for a short time about 1811, with his brother-in-law, Elisha Cheney, clock manufacturer, who for a time made screws and pins for the North pistols. Records indicate that all North con- tracts were signed solely by Simeon North. It is also in 1811, while engaged in fulfilling the contract of 1810, that North was commissioned as Lieutenant Colonel in the 6th Connecticut Regiment.

Shortly before the declaration of War of 1812, Colonel North, on instance of the Secretary of War, backed by promise of additional pistol contracts, enlarged the capacity of his Berlin factory's output by additional buildings and machinery. On April 16, 1813, the prom- ise was carried out. Col. North was given a contract for 20,000 pistols, and erected a new, large factory at Middletown, Connecticut, six miles from Berlin, where the new Model 1813 was manufactured, and doubtless many pistols of the 1810 contract completed. The Berlin shops were continued in operation under supervision of North's oldest son, Reuben, making forgings for the Middletown factory, until 1843, when they were closed. The Berlin factory was destroyed by flood in 1857.

In the pistols made at the Middletown factory, which was built along the most advanced ideas of the time, Colonel North successfully embodied the principle of standardization and interchangeability of parts, then a novel idea and subject to skeptical criticism.

The delivery under the contract of 1813, which was to be completed in five years, was delayed by a year, due to modifications introduced in 1816. July 1, 1819, before completing the delivery under the old contract, North entered into a new contract for 20,000 horse pistols (Model 1819) which he completed in 1823, well in advance of the specifications.

On November 16, 1826, Colonel North was awarded a contract for 1,000 navy pistols, (Model 1826). This contract was repeated by another for the same model and the same number Dec. 12, 1827, and again for an additional thousand on Aug. 12, 1828. These three thousand pistols, of Model 1826, were the last of the North pistol contracts. Colonel North had entered into the manufacture of rifles in 1823, and with the expiration of the 1828 pistol contract, he turned his entire attention to the manufacture of rifles, both standard muzzle-loading types and the Hall breech-loaders.   Between 1799 and 1828 Colonel North had manufactured and delivered approximately 52,000 pistols to the United States Government.

It is interesting to note that even the early date of the Republic were not free from economic difficulties and rising costs.   Though the North factories were for those days, organized along advanced lines, and had been in operation for many years, the pistols of Model 1819 cost the government $8.00 each as against $6.50 of 1799, and $6.00 of 1800.

U. S. FLINTLOCK PISTOL MODEL 1799, NORTH & CHENEY, ARMY, NAVY
Illustrated—Fig. 1, Plate 1.

Caliber .69, taking an ounce spherical ball. Eight-and-one-half inch round, smoothbore barrel.   Total length fourteen-and-one-half inches. Weight 3 pounds, 4 ounces. Brass mountings.   The pistol has no fore-end, and consists of a brass frame which contains the usual inner and outer flintlock mechanism, and into which the barrel is seated. The pan is cast integral with the frame or body of the pistol.   The one piece walnut grip is reinforced with an iron back-strap, let in flush, and extending from the breech screw and tang on top, to the brass butt cap, and held by a screw passing through the stock to the frame extension.   The brass trigger guard is fastened to the frame by two screws.   A steel, button-head ramrod fits into the frame on the lower right side.

The arm was made after the pattern of the French Model 1777 army pistol, and varies only in a barrel longer by one inch, a rounded breech underside at its seat in the frame, and in having an additional screw at the front end to hold the barrel more securely. The pistol was made without a belt hook.

The specimen described is marked on the barrel above the breech "V", "P" and "US", in three lines reading vertically from the muzzle. The frame is marked underneath, "NORTH & CHENEY BERLIN", in a curve. Specimens are also known marked "S. NORTH & E. CHENEY BERLIN". Others have 8-9/16 inch unmarked barrels.

The pistol was made at the Simeon North factory at Berlin, Conn., under a contract of March 9, 1799, for 500 pistols at $6.50 each, to be delivered in one year. Prior to the completion and delivery of these first 500 pistols, on Feb. 6, 1800, an additional contract for 1,500 pistols at $6.00, was made between Simeon North and James Henry, Secretary of the Department of War, on behalf of the United States.

The original contract of March 9, 1799, for the first lot of 500 of these pistols is believed to have burned in a fire of the War Department Buildings, Nov. 8th, 1800. A large portion of the records, between the year 1800, and August 25, 1814, were also destroyed when the British burned most of the government buildings in Washington in the invasion of the War of 1812-1815.

Though the contract of Feb. 6, 1800, for the 1,500 additional pistols, and the correspondence in connection with it were signed by S. North alone; and though the North family history claims that North's association with

Elisha Cheney never reached the status of partnership; and that they were not associated in business until 1811 or so; the fact remains that pistols of this model were marked NORTH & CHENEY, and the following extract from the contract of Feb. 6, 1800, clearly indicates that these are the pistols, with barrels set in a brass frame, that were contracted for under its terms: —

Articles

Second—"except that the Bore or Calibre of the pistols is to be the same with that of the pattern Charleville musket, and that the Barrel and Ramrod are each to be one inch longer than the pattern: and also that the part of the Breech of the pistol which lies within the Brass may be formed round on the under part instead of being squared, but it is agreed that the two squares shall be retained on each side of the Breech as may be necessary to give a firm position to the barrels within the Brass work —"

This second contract was of two years duration, and though the delivery was due to be completed on Feb. 6, 1802, the contract was not discharged until Sept. 11, 1802. It is quite possible that these contracts were awarded as a result of action on the War Department recommendations to the House of Representatives, March 4, 1794, for the procurement of arms and military stores to include a thousand pistols.

The organization of the United States army of the period between 1790 and 1800, included in Dec. 1792, four troops of "Light Dragoons" to serve whenever order-ed as "dismounted" dragoons; — one troop to each of the four Sub-Legions of the Army. The "Military Establish-ment of the United States" after October, 1796, was to include two companies of "Light Dragoons", to serve on horse or on foot. While it is quite likely that the rank and file were armed with musketoons, pistols must have formed a large part of the equipment of these organizations.

U. S. FLINTLOCK PISTOL MODEL 1806, HARPERS FERRY, ARMY, NAVY

Illustrated—Fig. 2, Plate 1.

Caliber .54 taking a halfounce spherical ball. Ten andonesixteenth inch round smoothbore barrel. Total length sixteen inches. Weight 2 pounds, 9 ounces. Brass mountings. The barrel is key fastened to the stock and has a rib extension underneath, from the end of the half stock to the end of the muzzle. The rib carries a steel ramrod thimble. The ramrod of the specimen described is steel, swell tipped. A brass front sight is set in at the muzzle and steel open rear sight is set on the barrel tang. The butt is slightly fishtail shaped at the butt cap. The brass cap has a three inch extension on each side, set flush with the wood and reaching upward to reinforce the butt. The trigger guard forks at the rear into a graceful curl to complete the oval. The rear branch of the guard continues under the stock and ends within onequarter of an inch of the butt cap. Though origin ally the pistol as issued was without sights, many are found with brass bead or low knifeblade front sight dove tailed into the barrel, and some have a Vnotch steel open rear sight brazed on the tang. From the similarity of many of these sights it is probable that the addition was made by the Ordnance Department, or on some army post while the pistols were in the service. It is also quite likely that the original issue had wood ramrods.

The flat bevelled lock-plate is 4-3/4 inches long, 7/8 inch wide. The hammer is flat, bevelled, double-necked. The steel pan is horizontal, with fence to the rear, and is forged integral with the lock-plate. The narrow frizzen, 5/8 inch wide and 1-1/2 inches high, comes to a rounded point at the top, and at the bottom has a shoulder on each side to cover the pan. The inside of the lock-plate is stamped "IV".

In the specimen illustrated, the barrel is marked with proofmark "P", "US", and is numbered "23". The lock-plate is marked with a spread eagle and "US" under the pan and "HARPERS FERRY 1806" in three lines behind the hammer. The walnut half-stock is marked "C" on the left side and with indistinguishable inspector's initials in an oval.

Early ordnance reports of production at the Harpers Ferry Armory* show the following pistols manufactured at that armory in: —

        1806—8 pattern pistols.
        1807—2880 pistols.
        1808—1208 pistols.

Although the ordnance reports record only eight pistols made in 1806, it is believed that a somewhat larger number of lock-plates dated 1806 were made and assembled into pistols in 1807, as is evidenced by the number "23" marked on the barrel of this specimen. However, pistols of this model dated 1806 are extremely rare. All Model 1806 pistols were numbered serially on the barrel. On the barrels of some of these pistols, dated after 1806, the proofmark "P" is combined with an eagle head, in an oval.

* Report of Lt. Col. G. Bumford, Chief of Ordnance, Nov. 30, 1822, Table D, Document No. 236, American State Papers Class V, Military Affairs.

These pistols, though famous for their beautiful lines, balance and accuracy, were rather weakly constructed for military use.

The Harpers Ferry Armory was established in 1796. Washington, attracted by the ample water power facilities at Harpers Ferry, Virginia, located at the confluence of the Potomac and Shenandoah rivers, selected it for the site of one of the two Federal armories and arsenals authorized by Congress in the Act of April 2, 1794. Harpers Ferry was named for Robert Harper, who settled there in 1747 and established a ferry across the Potomac. The site consisted of 125 acres of land purchased from the Harper family. Though the construction of buildings and shops was begun in 1796, the first output is recorded in 1801 when 293 muskets were made. During its existence the armory averaged over 10,000 muskets and rifles annually, and about 75,000 small arms were kept in storage reserve.

The armory gained considerable public attention in 1859, through its capture for a day by a rabid abolutionist John Brown, who, with a party of nineteen others, unsuccessfully attempted to seize arms for the arming and revolt of negro slaves. The abortive attempt cost John Brown his life by execution.

At the time of Virginia's secession, Harpers Ferry Armory was garrisoned by Lt. Roger Jones, U. S. Army, and a detachment of 45 enlisted men. On the night of April 18, 1861, confronted with the imminent capture of the armory by an assembling large body of Virginia

NOTE: The data on the early production of the U. S. Harpers Ferry and Springfield Armories is available in the "American State Papers, Class V, Military Affairs", a government compilation of military affairs and contracts reported on to Congress, and originally documented in U. S. Serials for the period 1798 to 1834.

militia, Lt. Jones set fire to the arsenal and the armory, destroying over 20,000 stored small arms and as much public property as possible, and retreated across the Potomac. Some of the arms parts, equipment and machinery were salvaged by the Confederates and were used by them later in the assembly of Confederate arms.

U. S. FLINTLOCK PISTOL MODEL 1808, S. NORTH, NAVY
Illustrated—Figs. 3 & 4, Plate 1.

Caliber .64 taking an ounce spherical ball. Ten-and-one-eighth inch round, smoothbore, browned barrel without sights. Total length sixteen-and-one-quarter inches. Weight 2 pounds, 14 ounces. Brass mountings. The barrel, front of the trigger guard, thimble and trigger are pin fastened. The full length stock extends to within one-quarter of an inch of the muzzle. The curved-in, swelled butt is reinforced by an iron back-strap extending from the tang to a short butt cap extension. The umbrella-shaped, brass butt cap forms a ridge at the forward edge and comes to a point on each side. The brass trigger guard forks at the rear into a graceful curl to complete the oval. The rear branch continues under the stock and ends just short of the forward point of the butt cap. The sides of the forward end of the trigger guard are pinched in. The hammer is double-necked. No slot in jaw screw. The horizontal pan has a fence. Hickory swell-tipped ramrod. An iron hook on the left side is held by the rear side screw and is pinned into the side plate.

The barrel is not marked. The flat, bevelled lock-plate is marked with an eagle and "U. STATES" between the hammer and the frizzen spring, and "S. NORTH BER-LIN CON." vertically in three lines behind the hammer.

These navy boarding pistols, equipped with a hook for thrusting into sash or belt, were made by Simeon North at Berlin, Conn., on contract with the Navy Department dated June 30, 1808, for 1,000 pairs of pistols, at $11.75 per pair. The contract was extended on Dec. 4, 1810, by an order for additional 500 pairs at $12.00 per pair.

In fulfilling this contract, North applied what was in those days a revolutionary principle: — standardization of parts. To quote from Simeon North's letter to Robert Smith, Secretary of the Navy, dated Nov. 7th, 1808: —

"To make my contract for pistols advantageous to the United States and to myself I must go to a great propor-tion of the expense before I deliver any pistols. I find that by confining a workman to one particular limb of the pistol until he has made two thousand, I save at least one quarter of his labour, to what I should, provided I finished them by small quantities; and the work will be as much better as it is quicker made.—"

The method of assigning the production of identical parts to the same individual workman saved in time and labor as well as improved the quality. The resulting uni-formity of parts was doubtless responsible for Colonel North's idea of interchangeability of all parts, which he incorporated later into pistols made under the 1813 and 1816 contracts, as well as for the invention of the necessary machinery and tools to put this idea into actual practice.

These Model 1808 navy boarding pistols rendered valuable service in the War of 1812.

Pistols are also known similar to the North Model 1808, Navy, but marked only with eagle and "U.

STATES" on the lock-plate, which is a half-inch longer than the North, and the lock of heavier construction,— caliber also .64. The maker is unknown.

U. S. FLINTLOCK PISTOL, MODEL 1810, S. NORTH, ARMY
Illustrated—Figs. 1, 2 & 3, Plate 2.

Caliber .69, taking an ounce spherical ball. Eight-and-five-eighths inch *round* smoothbore barrel without sights. Total length fifteen inches. Weight 2 pounds, 11 ounces. Brass mountings. The barrel, the front of the trigger guard, the thimble and the trigger are pin fastened. The full length stock extends to within one-quarter of an inch of the muzzle. The curved in, swelled butt is reinforced by an iron back-strap extending from the tang to a short butt cap extension. The umbrella-shaped brass butt cap forms a ridge at the forward edge and comes to a point on each side. The brass trigger guard forks at the rear into a graceful curl to complete the oval. The rear branch continues under the stock, and ends just short of the forward point of the butt-cap. The sides of the forward trigger guard extension are pinched in. The hammer is double-necked. Horizontal brass pan with a fence. Hickory swell-tipped ramrod.

The barrel is marked with the proofmark "V" and an eagle head in an oval, with letters "CT" in the same oval, reading vertically from the muzzle. The flat bevelled lock-plate is marked with an eagle and "U. STATES" between the hammer and the frizzen spring, and "S. NORTH BERLIN CON." vertically in three lines behind the hammer.

This pistol was made by S. North at Berlin, Conn., under a contract for horse pistols made in 1810 with Tench Coxe, Purveyor of Public Supplies, and was similar in design to the Model 1808 Navy boarding pistols made by North, except that the caliber was .69, the barrel was shorter, and the pistol was made without a belt hook.

The contract itself, which was awarded under the provisions of the "Act of 1808 for the Arming and Equipping the Whole Militia" has not been located in the government files, probably having been burned when the British set fire to the public buildings in Washington, August 24, 1814, and the total output of this issue is unknown. Probabilities are that the original contract was for close to 5,000 pistols, as records indicate that 2,400 pistols were still due the government on this contract on April 16, 1813, the date of signing the contract for the later Model 1813 pistols.

The rarity of these pistols would seem to indicate that the 2,400 pistols undelivered by April, 1813, were most likely made according to the specifications of the new contract. Indeed specimens are known, made with the round barrel (throughout its length), and the brass mountings of this model, and the North Berlin lock-plate, but with the stronger double band construction of the 1813 model. Others have the simpler design and the iron mountings of the 1813, but with a North Berlin lock-plate and are marked on the barrel "P" and "US" in the manner of the 1813 pistols. These were doubtless assembled and completed in the new North factory erected at Middletown in the summer of 1813, and were partly equipped with parts previously manufactured at the Berlin factory, and partly with new parts made at Middletown.

These pistols are called Model 1799 in the North family memoir, "Simeon North First Official Pistol Maker," written by Simeon North's descendants. However study of a letter written in connection with the contract of May 30, 1808, casts considerable doubt on that classification. This letter written by Simeon North to the government suggesting certain modifications, reads in part as follows: —

> "I have this day received a pattern pistol from the Hon. Joseph Hull, Esq. of Derby. It being a large navy pistol nearly a third larger than a common horseman's pistol. I find on examining the said pistol there may be some alterations which may be much to the advantage of the United States, which is as follows (viz) A strap of iron running from the britch pin to the cap, which would be a very great support to the stock, also the pans to be made of brass instead of iron, some other small alterations in the different parts might be made to the advantage of the pistol, all of which I would humbly submit to your honour."

North's proposals were accepted, and the improved pistols contracted for and made. That arm is illustrated in the North book as the Model 1808 navy pistol. Comparison of this arm with the pistol called Model of 1799, shows that it is alike in all its features except barrel length, total length and the belt hook. It seems reasonable to believe that if North had made such an arm in 1799-1800, he would not have suggested its certain characteristic features as new ideas or improvements in 1808, but by reference to the other pistol, would have merely suggested an addition in length and a belt hook.

U. S. FLINTLOCK PISTOL, MODEL 1813, S. NORTH, ARMY, NAVY

Illustrated—Figs. 1 & 2, Plate 3.

Caliber .69, taking an ounce spherical ball. Nine-and-one-sixteenth inch round smoothbore barrel with *semi-octagonal breech*. Total length fifteen-and-one-quarter inches. Weight 3 pounds, 6 ounces. Iron mountings, double strap banded barrel without sights. The band is spring fastened by a stud fitting between the straps. The *front strap* of the band *is not fluted*. The curved-in, swell shaped, rounded butt is reinforced by an iron back-strap running from the tang to the iron butt cap. The stock ends flush with the front of the barrel band. The lock-plate is flat and bevelled at the front and is slightly rounded at the rear. The pan is brass, tilting up at the rear and has no fence. The frizzen tilts backwards towards the hammer. The swell tipped hickory ramrod ends in a slotted iron ferrule, threaded inside to take a wiper head or a bullet screw. Pistols of this model were also made for navy issue, and carried a hook on the left side, fastened to the rear side screw, for hooking into a sash or belt.

The barrel is marked with the usual proofmark "P" and "US". The lock-plate is marked "S. NORTH", "US" and "MIDLn CON.", between the hammer and the pan. In the specimen illustrated, the usual eagle is missing, or is indistinguishable as a result of light stamping and wear. A few specimens are known marked only "S. NORTH US" in two lines between the hammer and frizzen spring.

These pistols are quite rare. The original government contract with Simeon North made April 16, 1813, was for 20,000 pistols: — however, relatively few were made of this model and caliber. These first, caliber .69 issues, of which only 1,150 were delivered by June 22, 1815, proved unpopular due to excessive kick, and were the subject of an adverse report from Colonel Decius Wadsworth of the Ordnance Bureau to the Secretary of War on June 10, 1815. Colonel Wadsworth wrote: — "the caliber of the pistols for greater simplicity might be the same as that of the rifle. It is essentially wrong in my opinion, to give a pistol the caliber of a musket, which I am informed, has been done in some of those made for the United States service."

As a result of this criticism the first issues of the large .69 caliber, using an ounce ball, which were manufactured in 1814 and 1815, were discontinued and superceded by the smaller caliber, .54 Model 1816 North pistols which came within the same contract.

The original contract for the 20,000 pistols called for completion of deliveries within five years at a price of $7.00 per pistol, which included ten bullet screws, ten screw drivers and a stated number of replacement parts, such as springs, pins, and external lock-parts, per 100 pistols. The contract further specified that the weight was not to exceed 3-1/2 pounds and that "the component parts of the pistols were to correspond so exactly that any limb or part of any one pistol may be fitted to any other pistol of the 20,000."

This contract of 1813 was the first in which a contractor agreed to a stipulation to produce an arm uniform and interchangeable in its parts, and marked a revolutionary change in the manufacture of firearms, though the

system of interchangeable parts had had a limited application in the manufacture of arms of other armorers prior
to this date.

Model 1813 pistols were issued for service in the
Seminole War.

U. S. FLINTLOCK PISTOL, MODEL 1816, S. NORTH, ARMY

Illustrated—Fig. 3, Plate 3.

Caliber .54, taking a half-ounce spherical ball. Nine
inch round, smoothbore barrel. Total length fifteen-and-
one-quarter inches. Weight 3 pounds, 3 ounces. The
barrel and the iron mountings were acid browned, the
lock case-hardened. A brass knife-blade front sight is set
on the fore strap of the double strap barrel band, which is
fastened on the right side by a stud fitting between the
straps. The curved-in, swell shaped rounded butt is reinforced by an iron strap running from the tang to the iron
butt cap. The lock-plate is flat, bevelled in front, rounded
at the rear. The pan is brass, tilting up at the rear, and
is without fence. The frizzen tilts backwards, towards
the hammer. The swell-tipped hickory ramrod ends in a
slotted iron ferrule, threaded inside to take a wiper head
or a ball screw.

In the specimen illustrated the barrel is marked with
the usual proofmark "P" and letters "JN"; probably those
of the barrel maker, or possibly of the proof tester. The
lock-plate is marked "S. NORTH", an eagle, "US" and
MIDLn CON.", between the hammer and the pan.

The large .69 caliber of the original Model 1813 pistol

having proved unsatisfactory, the original contract of April 16, 1813, was changed and modified in 1816, at an increased cost of one dollar per pistol; the five years dura-tion of the contract being extended for one year. Its new provisions included the reduction of the caliber from .69 to .54; the inclusion of a sight; the extension of the princi-ple of interchangeability of parts to the mechanism of the lock; the browning of the barrel and the mountings, and the case-hardening of the lock. Evidently the system of interchangeable parts of the lock had not been included in the original contract. Other differences are: — the barrel is rounded throughout its length, and the tip of the stock projects beyond the end of the barrel band. Of the 20,000 pistols contracted for in 1813, the vast majority of those manufactured were of this model.

This model was issued for service in the Black Hawk, Seminole and the Mexican Wars. Many were converted to percussion in 1850.

U. S. FLINTLOCK PISTOL, MODEL 1816, S. NORTH, ARMY, LATE VARIATION.

Caliber .54 taking a half ounce spherical ball. Nine inch round smoothbore barrel. Total length fifteen-and-one-quarter inches. Weight 3 pounds, 2 ounces. The barrel and the iron mountings were acid browned. A brass knife-blade front sight is set on the fore-strap of the barrel band which is spring fastened by a stud on the right side, between the two straps of the band. The curved-in swell-shaped, rounded butt is reinforced by a long iron butt-strap extending from the tang to the butt cap. The pan is brass, tilting up at the rear and without fence. The frizzen tilts backward toward the hammer. The swell-tipped hickory ramrod ends in a slotted iron ferrule threaded inside to take a bullet screw or a wiper.

In the specimen described, and similar in all outward appearance to the regular Model 1816, the barrel is marked with the usual proofmark "P" and "US". The lockplate is marked "S. NORTH", an eagle, "US" and "MIDLtn CONN." The stock is marked with the government inspector's initials "LS", in an oval. All mountings are stamped "B".

This model varies from the regular 1816 model only in the stamping of the lock-plate. Like the Model 1816, it comes within the contract for 20,000 pistols placed with Simeon North, April 16, 1813, the deliveries to be completed in five years, and extended an additional year by the changes to caliber .54, approved May 9, 1816. Some of the last of the pistols to be delivered under this contract were manufactured concurrently with the new Model 1819 pistol, contracted for by Col. North on July 1, 1819, prior to the completion of the deliveries of all the 20,000 pistols under the contract of 1813. These last Model 1816 pistols to be delivered were stamped with the new S. North die of the type used on the new 1819 models. The lettering is smaller than on the regular 1816 pistol; the design of the eagle slightly different, and the spelling has been changed from "MIDLn CON", on the regular model, to "MIDLtn CONN."

This model, like the earlier 1816, saw service in the Black Hawk, Seminole and Mexican Wars.

U. S. FLINTLOCK PISTOL, MODEL 1818, SPRINGFIELD
ARMORY, ARMY.

Illustrated—Fig. 4, Plate 3 & Fig. 1, Plate 4.

Caliber .69 taking an ounce spherical ball. Eleven-and-one-sixteenth inch round smoothbore barrel. Total length seventeen-and-three-quarter inches. Weight 3 pounds, 3 ounces. The barrel is held by a double-strap iron band with a brass front sight on the front strap. The band is spring fastened by a stud fitting between the straps. Iron trigger guard, back-strap and butt cap. Bevelled edge lock-plate 5-1/8 inches long, 1-1/16 inches wide. Horizontal iron pan forged integral with the lock-plate. The hammer is of the old fashioned goose-neck type, flat, with bevelled edges. The hammer screw has only a slot in the top, without the usual transverse hole. The frizzen is similar to the Charleville, with turned up toe, and is ¾ inch wide, 1-1/2 inches high. The frizzen spring ends in a diamond shaped point. The hickory ramrod is swell tipped, and fitted at the small end with a slotted iron ferrule, threaded inside to take a wiper head or a ball screw.

In the specimen illustrated the barrel is marked "P" in an oval, and "V" with an eagle head between the two letters, which read from muzzle to the breech in a vertical line. The barrel is dated over the breech-plug "1818". The guard-plate is marked "EA" and the underside of the barrel is marked "EA7."

The lock-plate is marked "US" under an eagle, between the hammer and the frizzen spring, and "SPRING-FIELD 1818" in three lines in rear of the hammer; the word Springfield being in the two upper lines. The rear of the lock-plate is similar to that of the Charleville musket, sloping down to a tit-like point. The inside of the lock-plate is marked "LD" at the top and "B" in the center. Of the outer lock parts, the frizzen, hammer, hammer screw, and all the screw heads are marked "X". All the inner lock parts are marked with three parallel file cuts. The long, black walnut stock slopes to a swell-shaped rounded butt, is oil finished, and is marked on the left side with inspector's initials "S" and "ET" stamped in script. Specimens are also known marked inside the lock-plate "S. DALE" and "J. CROSBY".

The pistol is quite rare. Records indicate that one thousand only were made in 1818 at the Springfield Armory from major parts on hand, the assembly of the arms having been authorized by the Ordnance Office in March, 1817. The pistols were all of the same model, but were fitted with two different types of locks.

It is believed that due to the shortage of pistol locks in the government and contractor armories, and the desire to maintain the increased production of muskets in the Springfield Armory: — (13,015 muskets were made in 1817 as against 7,199 in 1816); the first five hundred pistols of this model were equipped with imported English or Belgian locks. The other five hundred pistols of this issue were fitted with the usual double-necked hammer of the U. S. military models, and it is an unsettled question whether they were also made abroad, and fitted with American hammers at the time of assembly; or whether the entire second lot were made in United States. A possible confirmation of the latter theory is the fact that the goose-necked locks are equipped wih a prong-retained main spring, a type of fastening more popular in Europe than in United States. Again, in the double-necked hammer models, the terminal point at the rear of the lock-plate is slightly different, and resembles more the usual North and other pistols of the period.

It is difficult to understand why caliber .69, which was found unsatisfactory in the Model 1813 North pistols and was discontinued in 1816, was re-introduced for manufacture and issue in 1818, unless the specifications for this arm were drawn up, pattern pistols made and the barrels forged prior to 1815. Indeed, confirming this theory, a few specimens of this model are known clearly marked on the lock-plate with the date "1815".

Apparently the entire issue was experimental and this model, clumsy, ill-balanced and unnecessarily large in both size and caliber, proved unsatisfactory even as a utilitarian weapon made purely for combat service; for after the first thousand no more were made.

It will be noted that the dating of this pistol as Model 1818 is based on the production year, and is merely a temporary expedient. In view of the existence of locks dated 1815, it is obvious that the date of the *approval* of the *pattern* by the War Department, or of the contract for the foreign locks, antedates even that earlier year. It is hoped that additional data will be discovered which will permit the assignment to this arm of an accurate model date.

The Springfield Armory developed gradually out of an arsenal and powder magazine established on Washington's approval at Springfield, Mass., in 1777. Begun at first as a depot for the manufacture of musket cartridges and gun carriages, the depot soon broadened its activities to the repair of small arms and the preparation and supply of munitions or war and ordnance of all kinds to the Continental armies. When in 1792 Congress authorized the establishment of two national arsenals, and on April 12, 1794, directed the construction of two Federal armories, President Washington combined the storage and manufacturing authorizations and selected for the combined purposes, Springfield in the North, and Harpers Ferry in the South. The manufacture of arms at the Springfield Armory began in 1795, in which year 245 muskets were produced, mostly filed laboriously by hand. Production facilities were rapidly increased until by 1825, the armory reached an annual output of 15,000 muskets.

The early pistol production record of the Springfield Armory is confined to:

    1818—1000 pistols
    1856—2710 pistol-carbines (Model 1855)
    1857—1311 pistol-carbines

Although not indicated in the early production tables, pattern pistols, such as of Model 1842, are known to have been made at the armory.

Under the able management of the U. S. Ordnance Department the Springfield Armory maintained a splendid record of service in all national emergencies. It reached its peak production in October, 1918, during the World War, with a daily output of well over one thousand rifles.

U. S. FLINTLOCK PISTOL, MODEL 1819, S. NORTH, ARMY
Illustrated—Fig. 2, Plate 4.

Caliber .54, taking a half ounce spherical ball. Ten inch round, smoothbore barrel. Total length fifteen-and-one-half inches. Weight 2 pounds, 10 ounces. The barrel and the iron mountings were acid browned. The barrel is held by a spring-fastened, single band formed like the lower band of a musket, and carries a brass knife-blade front sight on the muzzle. The barrel tang carrying a large, open rear sight extends down to the rounded, swell-shaped curved-in butt to meet a short branch of the butt plate cap. The lock-plate is flat, bevelled in front, rounded at the rear. The pistol is provided with a swivel ramrod and a sliding safety bolt behind the hammer, on the exterior of the lock, to hold the hammer at safety at half cock. The pan is brass, without fence.

In the specimen illustrated the barrel is marked with proofmark "P", "US" and "X". The lock-plate is marked "S. NORTH", an eagle, "US" and "MIDLtn CONN." between the hammer and the frizzen spring, and is dated 1822 behind the hammer.

This pistol with its longer and slender barrel was better balanced and more symmetrical in appearance than its predecessor, Model 1816. The additional inch of barrel length resulted in a longer period of gas pressure and improved accuracy. Designed for mounted service as a horsepistol to be carried in saddle holsters, the swivel ramrod facilitated loading and eliminated the possibility of accidental dropping or loss, incidental to the old type wooden ramrod. However, the safety bolt which was meant to prevent accidental discharge when a pistol was drawn from the holster, proved more of a hindrance than an aid, and was discontinued in all later models.

This model is based on patterns furnished Simeon North by the Ordnance Bureau May 21, 1819. Probabilities are that the pattern pistols for it were made at the Harpers Ferry Armory which according to ordnance reports, made six pattern pistols in 1819.

The contract for this model was signed July 21, 1819, before Colonel North had completed the delivery of all the 20,000 pistols he had undertaken to supply to the War Department under the contract of 1813. The new contract was also for 20,000 pistols, (10,000 pairs), extending over five years, and deliverable at the rate of 4,000 each year. The contract price was $8.00 per pistol.

By this time the North factory was organized for large scale production. Col. North was able to complete the contract fourteen months ahead of time, 2,000 pistols being delivered within the first year, in 1820; 7,000 in 1821, 8,000 in 1822, and the last 3,000 in 1823.

These pistols were used in the Black Hawk, Seminole and Mexican Wars.

U. S. FLINTLOCK PISTOL, MODEL 1826, S. NORTH, ARMY, NAVY

Illustrated—Fig. 3, Plate 4.

Caliber .54 using a half ounce spherical ball. Eight-and-five-eighths inch round smoothbore barrel. Total length thirteen-and-one-quarter inches. Weight 2 pounds, 4 ounces. The barrel and the iron mountings were acid browned. The barrel is held by a spring fastened, single band, formed like the lower band of a musket, and carries a brass, knife-blade sight, at the muzzle. The iron barrel tang carrying a large open rear sight, extends down to the rounded butt to meet a short extension of the butt plate cap. The bend of the full length stock forms almost a right angle, and the front extends up to the swivel ram-rod. The lock-plate is flat and bevelled in front, and the surface is slightly rounded at the rear. There is no safety bolt. The brass pan is tilted upward at the rear and is without fence. Double-necked hammer.

In the specimen illustrated, the barrel is marked "P", "US" and "ET". The lock-plate is marked "US" and "S. NORTH", in two lines between the hammer and the frizzen spring, and is dated 1828 behind the hammer.

This model was also made with belt hook on the left side of the stock, for navy use, and some were made with tinned barrels. In those made for sea service and equipped with a belt hook, the hook was held by one separate screw (not the rear side screw) and a pin.

These pistols were made by Col. Simeon North at Middletown, Conn., under a contract for 1,000 at $7.00 each, dated Nov. 16, 1826, for delivery within one year. Two additional contracts, also for 1,000 each, were made Dec. 12, 1827, and Aug. 8, 1828, and were carried out.

These 3,000 pistols were the last of the North con-tract pistols. After 1828 Col. North turned his entire attention to the manufacture of rifles for the government. The rifles were muzzle-loaders and Hall breech-loading rifles and carbines.

U. S. FLINTLOCK PISTOL, MODEL 1826. W. L. EVANS. NAVY
Illustrated—Fig. 4, Plate 4.

Caliber .54, using a half ounce spherical ball. Eight-and-five-eighths inch round smoothbore barrel. Total length thirteen-and-three-eighths inches. Weight 2 pounds, 4 ounces. Iron mountings. The barrel is held by a spring fastened, single band formed like the lower band of a musket, and carries a brass knife-blade front sight at the muzzle. The barrel and mountings were browned. The iron barrel tang carrying a large open rear sight, extends down to the rounded butt, to meet a short extension of the butt cap. The bend of the full length stock forms almost a right angle, and the front extends up to the swivel ramrod. The lock-plate is flat and bevelled at the front, and the surface is slightly rounded at the rear. The brass pan is tilted upward at the rear, and is without a fence. Double-necked hammer. The pistol is equipped with a steel hook, four-and-five-eighths inches long, on the left side of the stock, fastened by a separate screw (not the rear side screw), and pinned in place. The small end of the swivel ramrod is threaded inside to take a ball screw or a wiper head.

In the specimen described the barrel is marked "P". The lock-plate is marked with an eagle head, "W.L. EVANS" and "V.FORGE" between the hammer and the frizzen spring, and "1831", "U.S.N." behind the hammer.

The pistol was made on contract by W. L. Evans at Valley Forge, Pa., and is similar in all respects to the S. North Model 1826 pistol; most parts being interchangeable. Probabilities are that a North 1826 pistol was used as a model.

Some of these pistols are found dated 1830 behind the hammer and are marked "U.S. W.L. EVANS" between the hammer and the frizzen spring, and marked on the barrel "US", inspector's initials, and some are proof-marked "P" in addition. Probabilities are that this type of marking was on the earlier issues, and was changed at the request of the Navy Department to show the con-tractor's address and the "USN" replaced "US" to indi-cate Navy Department property. The iron parts of some of these navy pistols were tin plated to prevent or mini-mize corrosion while on sea service.

The Valley Forge was operated by the Potts family of Chester Co., Pa., as a general manufactory of iron products from 1757, until its destruction by the British under Gen. Howe, Sept. 21, 1777, about two months before Washington selected and occupied Valley Forge as his winter encampment. After the War the forge was rebuilt, and in conjunction with a slitting mill erected by Isaac and David Potts, resumed operations, manufactur-ing saws and domestic and farm implements. In 1786 the forge and the mill were operated by Isaac Potts and his son, James.

In 1814 the works were sold to John Rogers, a Phila-delphia hardware merchant. Early in 1821 Brooke Evans and John Rogers together converted the shops into an arms factory, the armory being known as Valley Forge. Brooke Evans and John Rogers on March 21, 1821, took over a defaulted contract awarded to Alexander McRae, of Richmond, Va., on July 28, 1817, for 10,000 muskets. Apparently after the contract was fulfilled the partnership was dissolved, for on January 1, 1825, Rogers alone obtained a contract for 5,000 muskets. This contract was probably shared with William L. Evans, of Evansburg, a practical gunmaker who managed the works.

In addition to the musket and above pistol contract, William L. Evans obtained a contract for 1,500 muskets in January, 1832. Some of the early arms made at Valley Forge are marked "V. FORGE", others "B. EVANS VALLEY FORGE", while the later pieces, made after 1825, are marked "W. L. EVANS V. FORGE".

William L. Evans was born May 28, 1797, at Evansburg, Montgomery Co., Pa., about five miles due north of Valley Forge. He was the 6th child of Owen Evans, who made guns for Pennsylvania in 1797, and in 1808 had a contract with the U. S. for making muskets. Owen was associated with Oliver Evans, of Philadelphia, manufacturer of flour-mill machinery, in connection with the Pittsburg Steam Mill, from about 1809 until his death in 1812.

William L. Evans lived at the family homestead from about 1829 to 1839, and there is a tradition that a factory stood in back of the residence. In 1825 he became associated with John Rodgers, who had purchased Valley Forge in 1814. It is believed, however, that barrels mostly, were made at Valley Forge, and the rest of the gun made in the factory at Evansburg, or assembled in the homes of the various workmen, as was frequently the case in those days, and specimens of this model are known marked "W. L. EVANS E. BURG" on the lock-plate. The Valley Forge gun factory was partially destroyed by a freshet in 1839, and was completely destroyed in 1843.

William Evans later sold the homestead to his bachelor brother Edward Evans, the postmaster, and removed to a place about one-half mile south of Evansburg, on the west side of Skippack Creek, where he died August 6, 1861. He is buried in the family plot in the old church-yard of the St. James Perkiomen Church, at Evansburg.

U. S. FLINTLOCK PISTOL, MODEL 1836, R. JOHNSON, ARMY
Illustrated—Fig. 1, Plate 5.

Caliber .54 taking a half ounce spherical ball and a charge of 50 grains of black rifle powder. Eight-and-one-half inch round, smoothbore, bright barrel. Total length fourteen inches. Weight 2 pounds, 10 ounces. All mount-ings are iron, polished bright. The lock-plate, hammer and frizzen were case-hardened; the frizzen spring, barrel tang, trigger and screws were blued. The barrel carries a brass knife-blade front sight, and is held to the stock by a single branch-band fastened on the left side through the forward side screw. A large open rear sight is on the tang. A long butt cap extension forms the back-strap. The three-quarter length, oil finished black walnut stock ends short of the swivel ramrod, which is threaded inside at the small end to take a wiper head or a ball screw. The hammer is double-necked. The brass pan has a fence, and is tilted upward slightly at the rear.

In the specimen illustrated the barrel is marked "US", "P" and "J.H.". The flat bevelled lock-plate is marked with "US", "R. JOHNSON MIDDN CONN." and dated "1841". All lock parts are stamped with number 77.

This pistol was made on contract with R. Johnson of Middletown, Conn., dated March 14, 1840, for 15,000 pistols at $7.50 each, to be delivered within five years, 3,000 pistols annually. R. Johnson had been previously awarded a contract for 3,000 of these Model 1836 pistols on June 27, 1836, at $9.00 each, to be delivered by June 1, 1837.

This model is the last of the U. S. martial flintlock pistols, and was made until 1844. These pistols were also made by the Waters Armory at Millbury, Mass., and were well thought of for their excellent workmanship, fine lines and good balance. These arms were used in the Mexican War. Many were converted to percussion in 1850.

Robert Johnson operated an armory at Middletown, Conn., from 1822 to 1854. In addition to the above pistol contracts he made long arms on the following contracts:

December 10, 1823—3,000 rifles.

July, 1829—600 Hall breech-loading rifles. (This contract was later changed to delivery of muskets.)

U. S. FLINTLOCK PISTOL, MODEL 1836. A. WATERS, ARMY
Illustration same as for R. Johnson, above.

Caliber .54 taking a half ounce spherical ball and a charge of 50 grains of rifle powder. Eight-and-one-half inch round, smoothbore, bright barrel. Total length fourteen inches. Weight 2 pounds, 10 ounces. All mountings are iron, polished bright. The lock-plate, hammer and frizzen were case-hardened; the frizzen spring, barrel tang, trigger and screws were blued. The barrel carries a brass knife-blade front sight, and is held to the stock by a single branch-band fastened on the left side through the forward side screw. A large open rear sight is on the tang. A long butt cap extension forms the back-strap. The three-quarter length, oil finished, black walnut stock, ends short of the swivel ramrod, which was threaded inside at the small end to take a wiper head or ball screw. The hammer is double-necked. The brass pan has a fence and tilts slightly upward at the rear.

In the specimen described the barrel is marked "US", "P" and "JH". The flat bevelled lock-plate is marked with an eagle head, and "A. WATERS, MILBURY, Ms. 1838".

This pistol was made on contract with Asa Waters, of Milbury, Mass., dated Sept. 22, 1836, for 4,000 pistols at $9.00 each.

This model is the last of the U. S. flintlock martial pistols, and was made until 1844. Pistols of this model were also made by R. Johnson at Middletown, Conn., and were favorably known for their fine lines, good balance and excellent workmanship. These pistols were used in the Mexican War. Many were converted to percussion in 1850.

The Waters Armory was founded in Millbury, Mass., by Asa and Elijah Waters, sons of a Revolutionary War gunsmith, Asa Waters, (1742-1813). The brothers learned the trade at their father's Sutton Waters Armory, on Singletary Stream, Sutton, Mass. The Sutton Waters Armory furnished muskets to the Continental troops, and is reputed to have been the first to utilize water power in the manufacture of arms.

In 1797 Asa (Jr.) and Elijah Waters purchased land on Blackstone River, and in 1808 built the Waters

Armory. Elijah died in 1814 and Asa Waters became the sole proprietor. The welding of barrels at the Waters Armory was done by a water power operated trip hammer invented and patented October 25, 1817, by Asa Waters.

In addition to the above contract Asa Waters had earlier received the following contracts for long arms:

August 13, 1818 — 5,000 stands of arms.
October 16, 1818 — 10,000 stands of arms.
October 16, 1823 — 10,000 stands of arms.

U. S. FLINTLOCK PISTOL, MODEL 1836, A. H. WATERS & CO. ARMY

These pistols are identical in all respects with the Model 1836 A. WATERS pistol described above, except for the marking on the lock-plate, which is stamped "A. H. WATERS & CO. MILBURY, MASS." and dated 1844.

The pistols were made on contract with A. Waters & Son, Milbury, Mass., (see A. WATERS pistol above). The contract was awarded Feb. 7, 1840, for 15,000 flint-lock pistols at $7.50 each, to be delivered over a period of five years, at the rate of 3,000 annually.

Most of the pistols made under this contract were marked "A. WATERS". Apparently in 1844 Asa H. Waters took over the management of the plant and incor-porated, as pistols dated 1844 are stamped "A. H. WATERS & CO."

Upon disposal of this contract, the Waters Armory

machinery was sold to William Glaze of Columbia, South Carolina, and was later used to make the Palmetto Armory arms.

In 1850 many of these Model 1836 flintlock pistols were converted to the percussion system, usually by any one of the three government authorized alterations. (See next chapter.) However, a few were convertd to per-cussion by use of a tape-primer device, such as Gedney, Maynard or Ward. (Illustrated—Figs. 2 and 3, Plate 5)

BUTT CAP   BAND   BAND
S. NORTH 1808   S.NORTH M. 1813   S.NORTH M.1816

U. S. MARTIAL PERCUSSION PISTOLS

THE DEVELOPMENT OF THE PERCUSSION SYSTEM —
INCENDIARY MIXTURES — GUNPOWDER — FIRST CARTRIDGES — FULMINATES — FORSYTH LOCK — U. S. CONVERSIONS — MAYNARD TAPE PRIMER —

U. S. PERCUSSION PISTOLS —

Model 1842—H. ASTON.

Model 1842—H. ASTON & Co.

Model 1842—I. N. JOHNSON.

Model 1842—PALMETTO ARMORY.

Model 1843—N. P. AMES.

Model 1843—DERINGER.

Model 1855—SPRINGFIELD Pistol Carbine.

HARPERS FERRY Pistol Carbine.

UNITED STATES MARTIAL PERCUSSION PISTOLS

THE DEVELOPMENT OF THE PERCUSSION SYSTEM

Inflammatory and incendiary mixtures of violently combustible character were known to, and are recorded to have been used by the Greeks as early as in the Fourth Century B. C. Probabilities are that they were used even earlier in the Orient. These early incendiary compositions contained only sulphur, pitch, charcoal, tow and incense. As centuries passed, the mixtures were improved by addition of resins, salts, oakum, naphta, petroleum and nitre, and because of their highly inflammable character, these mixtures were used effectively in both offensive and defensive military operations, and in naval engagements to deal with personnel and combustible materiel such as ships, sails and engines of war. These mixtures did not however utilize the principle of expanding gases to throw a projectile.

It was not until the discovery of the properties of pure saltpetre, or potassium nitrate, some time in the second quarter of the 13th Century, that actual gunpowder came into use. To whom the actual honor of the discovery should be given may never be known. Certain it is however, that Friar Roger Bacon, of England, gave the formula, concealed in an anagram embodied in his famous essay, "De Mirabili Potestate Artes et Naturae", in 1242.

The discovery of the properties of gunpowder led to the development of the means of its utilization. It is definitely known that gunpowder was used as a projectile propellant at the Battle of Crecy in 1346, and crude cannon are shown in pictorial records some twenty years earlier, in the Christ Church folio Number 19, dated 1326, library of Oxford, England. Entries listing shipment of guns and powder to England in 1314 are found in the Town Records of the City of Ghent, Flanders (now Belgium).

The invention of gunpowder spelled the doom of knighthood and privilege, and was the beginning of democracy. Firearms placed in the hands of the helpless serf, a weapon that stripped the advantage of costly protective armor from the powerful nobles, who had been accustomed to rule roughshod over their helpless peasantry. The comparative ease of manufacture of gunpowder and the crude early firearms, gave a chance of success to rebellion against tyranny and oppression; rebellion born of desperation, which was formerly easily crushed by mounted men within the virtually complete safety of their armor.

In efforts to retain their prerogatives, possession of firearms or gunpowder was prohibited to the commoners, and the possession of a flintlock, which permitted ambush as compared to the betraying matchlock, was for a time a death penalty offense even for a soldier of France under Louis XIV. But this was a law that like many others, could be promulgated but not enforced, and within a decade was repealed.

The development of gunpowder was not only the first step in man's emancipation from serfdom and battle for freedom; but also helped him in his conflict against nature and wilderness, though to be sure, the evolution of arms was not without its instances of regrettable misuse.

Gunpowder consists of a mixture of about 75% salt-petre, 15% charcoal, 10% sulphur by weight. Of these the charcoal is the combustible, the saltpetre is the oxidizer, or supporter of combustion due to its oxygen content, while sulphur is added as a neutralizer, to make the mixture safe. When ignited the mixture explodes, that is, burns with extreme rapidity, and instantly forms a great volume of highly heated gas, which in its effort to expand and escape the narrow confines of the chamber, propels the projectile out of the gun.

In the early gunpowders the various ingredients were ground fine into a dust-like powder and were then mixed. While transported to the theatre of operations, the mixture being formed of materials of different weights and densities, tended to separate into its component parts, and while the first charges out of the barrels in which the powder was carried, were weak, due to the top layer consisting chiefly of the light charcoal, the succeeding loads increased in strength, sometimes resulting in casualties to the gunners themselves due to burst guns. To avoid this danger and reduce the hazards of accidental explosion, for a long time artillerymen carried the various components separately, and mixed them just prior to use, and even then in many cases the piece was fired by a train of slow burning powder laid along the top of the cannon, the gunners taking to their heels until the piece fired or blew up. It was a hazardous occupation at best, and the dangers of the old time artillerymen were not lessened by

the presence of their infantry escort with lighted matchlocks in the proximity of the open powder casks.

This mealy powder had another disadvantage in the
need of careful handling. The powder adhered to the
gun barrel in loading, and when compacted by ramming,
tended to form lumps which burned slowly with considerable residue and loss of explosive effect.

These disadvantages were overcome to a degree in the
development of "corned" powder, in the first part of the
16th Century. The process consisted of the addition of
a small percentage of water to the mixed powder and then
squeezing the paste under pressure through sieves. The
resulting granules did not separate in transit, were not
affected by ramming, and burned evenly and more rapidly
due to larger air spaces, and with less loss of gas, which
with slower burning powders tended to escape partly
through the touch-hole. The granules were sifted for uniform size, depending on the size of the weapon, and were
polished and to a certain extent waterproofed by being
revolved in a keg with a small quantity of dry graphite.

Towards the third quarter of the 16th Century the
methods of loading firearms were further improved by the
invention of the cartridge, which at first was but a paper
cylinder containing a fixed charge of powder to facilitate
loading of small arms while mounted. The powder charge
for artillery was also put up in paper or linen bags. By
1590 the small-arms cartridge was further improved by
the inclusion of the ball, by binding it into the neck of
the paper cylinder. This complete cartridge facilitated the
loading of small-arms, and permitted greater rapidity of
fire, as well as the carrying of prepared and complete

rounds, both as a part of the soldier's equipment, and in form of readily issued reserve ammunition in the wagon train. The inventor of the cartridge is unknown, and its invention was quite likely the result of gradual develop- ment. It is known that the infantry of the armies of Gus- tavus Adolphus of Sweden, (1594-1632), used the paper cartridge, twelve rounds being carried by the soldier in a leather pouch, while twenty additional rounds per man were carried in the wagons of the combat train.

The flintlock whose development was described in Chapter 1, remained the standard military ignition system for nearly two centuries. However it still retained the drawbacks of the uncertainty of the flint striking sufficient sparks to ignite the priming charge; the comparatively long period required to service the piece for firing; the tendency to misfire in wet or inclement weather, and the flash and smoke of the priming preceding the discharge of the arm itself. These limitations, inconveniences and disadvantages led to the development of the percussion system.

The fact that a number of chemical substances, such as fulminates of mercury and silver, and mixtures contain- ing potassium chloride had detonating qualities; — that is, were capable of violent and instantaneous explosion when struck, vibrated or heated; — was known for some time before the end of the 18th Century, but it remained for a Scotch clergyman, Reverend Alexander Forsyth, to make practical use of this discovery, by the invention of a detonating lock, which he patented in 1807. A famous French scientist Berthollet, had been one of the earlier chemists to experiment in the use of mixtures containing potassium chlorate or one of the fulminates, with view to

finding a substitute for gunpowder; but though he discovered fulminate of silver in about 1778, he abandoned
his research on discovery that the compounds produced
with these ingredients were too powerful for the metal-
lurgical science of the period, and too dangerous and
unstable to mix or handle in large quantities. Forsyth, an
enthusiastic amateur chemist, made similar experiments as
early as in 1793; but he approached the problem from a
different angle, with the idea of utilization of the fulminates or other detonating compounds, as *priming* for
gunpowder, by exploding it by means of a blow struck by
a gun lock hammer. Forsyth's earliest attempts were to
put the fulminate into the pan along with the priming
powder and have the hammer strike it; but the resulting
explosion merely blew the powder out of the pan without
exploding the main charge in the breech. His subsequent
system consisted of a gun lock in which a minute quantity of fulminate of mercury was exploded by the blow
of a small firing pin or plunger, which was struck by a
hammer, similar to the conventional percussion type; the
fire being transmitted through a vent to the main charge.
An original and ingenious feature of the lock was a care-
fully made and fitted, manually operated magazine, containing a supply of detonating powder sufficient for twenty
or so shots. The Forsyth lock though effective was too
delicate and expensive for military use, and was followed
by the development of the pill lock in which the charge
was fired by a small pellet of detonating compound, sometimes enclosed in a paper cover or capsule. This system,
by 1816, was replaced by the invention of the copper percussion cap containing fulminate of mercury, which was
placed on a hollow cone or nipple, and when struck by

the hammer, flashed the fire of the explosion into the main charge in the breech of the arm.

Who the original inventor of the copper cap was is a moot question. The honor was claimed by celebrated British gunsmiths, Manton, Egg and Purdey, and by Colonel Hawker of England. In the United States the claim was advanced by Mr. Joshua Shaw, an English artist, resident of Philadelphia, who applied for a U. S. patent on percussion caps in 1814. Shaw's first experiments were with a steel tube containing the fulminate, with the idea of repriming the tube for continued use. His next experiments were in 1815, with a pewter cap to be thrown away after use. But even as a steel cap or tube was too rigid to explode a small charge of fulminate, the pewter cap was too weak, and at length in 1816 he developed a copper cap which was successful.

Though the practical and successful application of the percussion system was known as early as 1816, it was not until 1842 that the system was adopted by the U. S. Government to succeed the flintlock as a standard martial firearm ignition. The slowness and hesitation in adoption of percussion by the military authorities was due to the common fault of the manufacturers in overloading the caps and using a poor quality of copper, which resulted in numerous accidents. Other reasons were the large stocks of flintlock arms on hand and under contract, and last but not least, were the distrust of the system, traditional conservatism, and apprehension of the inability to obtain primers in campaigns in the wilderness.

As a matter of fact, though the percussion system was

formally adopted in 1842, the War with Mexico, (1846-1848), was fought largely with flintlock muskets, rifles and pistols.

In 1848 the flintlock arms in government armories were inspected, and serviceable weapons were ordered to be altered to the percussion system. The conversion to the percussion system consisted of removal of the pan, frizzen, frizzen spring and the flintlock hammer, which was replaced by a percussion type hammer. The hollow cone or nipple on which the fulminate cap was seated, was attached to the barrel in one of the three following methods authorized in government conversions:

1st—The cone was screwed into the barrel over the breech and to the right side and the touch-hole was plugged.

2nd—A cylinder lug containing a cone was screwed into the touch-hole.

3rd—A bolster lug was brazed to the flat on the right side of the barrel covering the touch-hole. The cone was screwed into the bolster. In alteration of muskets, another bolster method of conversion was to cut off the end of the barrel and fit a new breech with the cone bolster included as a part of the breech forging.

After the adoption of the percussion system by the services, inventors were not slow to devise a system of automatic feed of percussion primers to firearm locks, because the small primer cap was difficult to handle, especially mounted. Percussion primer mechanisms were invented by Maynard, Ward, Butterfield and Lawrence.

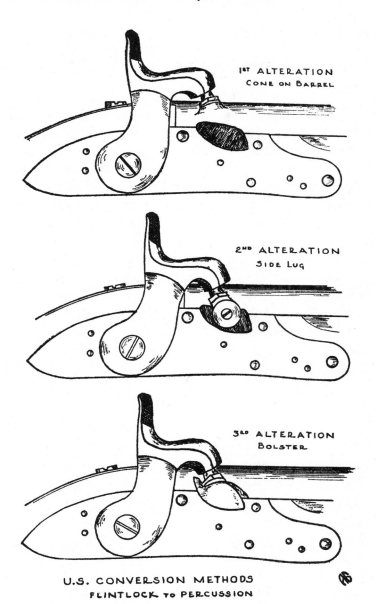

1ST ALTERATION
CONE ON BARREL

2ND ALTERATION
SIDE LUG

3RD ALTERATION
BOLSTER

U.S. CONVERSION METHODS
FLINTLOCK TO PERCUSSION

The general principle was similar, and since the Model 1855, Springfield pistol-carbine, described in this chapter used the Maynard system, this priming device will be described.

The Maynard tape primer was the invention of Edward Maynard, a dental surgeon of Washington, D. C. It consisted of a narrow strip of varnished paper of double thickness, having deposits of fulminating compound between the two strips, at equal distances apart. The strip was coiled in a recessed magazine in the lock-plate of the arm, and was pushed up by a toothed wheel when the hammer was cocked, so as to bring a fulminate cap on the top of the cone. The fall of the hammer exploded the cap, and at the same time, cut off the paper behind the exploded cap. Otherwise the functioning of the primer was the same as that of the copper percussion cap, the flash of the explosion being conducted through the hollow cone to the main charge in the breech and discharging the arm.

U. S. Percussion Pistol, MODEL 1842, H. ASTON, Army, Navy

Illustrated—Fig. 1, Plate 6.

Caliber .54. Eight-and-one-half inch round, smooth-bore barrel. Total length fourteen inches. Weight 2 pounds, 12 ounces. All mountings are brass. The barrel carries a brass knife-blade front sight, and is held to the stock by a single branch-band fastened on the left side through the forward side screw. The long butt cap extension forms the back-strap. The three-quarter length, black walnut stock is smoothly finished and ends short of the swivel ramrod, which is threaded inside the small end to take a wiper head or ball screw. All steel parts except the trigger, which is blued, are bright finish, burnished to a high polish. The pistol fired a half ounce spherical ball and took a fifty grain charge of black rifle powder.

The barrel of the specimen illustrated is marked on the barrel "US", "P" and the workman's initials "GP", as well as the date 1850 on the tang. The flat bevelled lock-plate is marked "US", "H ASTON" in front of the hammer, and "MIDDtn CONN. 1850" behind the hammer. Government inspector's initials "GW" and "WAT" in script are stamped on the left side of the stock.

The pistol was made by Henry Aston at Middletown, Conn., under contract of February 25, 1845, for 30,000 pistols at $6.50 each, the delivery to be completed within five years. Pistols of this model were also manufactured by Ira N. Johnson at Middletown, Conn., and the Palmetto Armory at Columbia, South Carolina. They were internationally known as the best made martial pistols of the time.

The firm of H. Aston, makers of the pistol, was organized about 1843 by Henry Aston who arrived in United States from England in 1819, and as a skilled pistol maker readily found employment with Simeon North, pistol manufacturer, at Middletown, Connecticut.

Subsequent to 1850, the firm of H. Aston was reorganized, and from 1851 the Aston pistols were marked "H. ASTON & CO."

The government pattern pistols of this model were made at the Springfield Armory. The government models were numbered on the barrel and marked "U. S. MODEL 1842", "V", "P" and an eagle's head. In addition to the usual markings "US", "SPRINGFIELD" and "1842", the larger parts were stamped "USM" for United States Model, while the smaller parts, such as screws, were marked "M". Though the original patterns were designated as Model 1842, the pistols were not manufactured until 1844, and were not issued to the services until 1845. The navy pistols of this model have an anchor stamped on the barrel.

U. S. PERCUSSION PISTOL, MODEL 1842, H. ASTON & CO., ARMY, NAVY

These pistols are identical in all respects with the Model 1842 H. ASTON pistol described above, except for the marking on the lock-plate which is stamped "US", "H. ASTON & CO.", "MIDDtn CONN. 1851".

Henry Aston's partners were Nelson Ashton, Peter Aston, John North, Sylvester C. Bailey and Ira N. Johnson.

U. S. PERCUSSION PISTOL, MODEL 1842, I. N. JOHNSON, ARMY, NAVY

Illustration as for H. Aston above.

Caliber .54. Eight-and-one-half inch round smooth-bore barrel. Total length fourteen inches. Weight 2 pounds, 12 ounces. All mountings are brass. The barrel carries a brass, knife-blade front sight and is held to the stock by a single branch-band fastened on the left side through the forward side screw. The long butt cap extension forms the back-strap. The three-quarters length black walnut stock is smoothly finished, and ends short of the swivel ramrod, which is threaded inside the small end to take a wiper head or ball screw. All steel parts, except the trigger which is blued, are finished bright, burnished to a high polish. The pistol fired a half-ounce spherical ball and took a fifty grain charge of black rifle powder.

The specimen described is marked on the barrel "US", proofmark "P" and initials "JH". The flat bevelled lock-plate is marked "US", and "I. N. JOHNSON" in front of the hammer and "MIDDtn CONN. 1853" behind the hammer. Government inspector's initials "JH" and "WAT" in script, appear on the left side of the stock.

The pistol was made by Ira N. Johnson under contract of March 28, 1851, for 10,000 pistols at $6.75 each. Ira N. Johnson had been one of the partners of H. Aston & Co., of Middletown, Conn. Having obtained the contract, Johnson severed his connection with the Aston Company and completed the contract alone.

U. S. PERCUSSION PISTOL, MODEL 1842, PALMETTO ARMORY, ARMY, NAVY

Illustration as for H. Aston.

This pistol was made under contract by William Glaze & Company at their Palmetto Armory at Columbia, South Carolina, and is identical in appearance and construction with the Model 1842 pistols made by Aston and Johnson at Middletown, Conn., and described above.

The barrel is marked "Wm. GLAZE & Co.", proof-marks "P" and "V", a palmetto tree and the date 1853 on the breech. The lock-plate is marked in front of the hammer "PALMETTO ARMORY", "S.C.", surrounding a palm tree. Back of the hammer the lock-plate is marked "COLUMBIA S.C.", and dated 1852. This is the rarest of the regular issue Model 1842 pistols.

The Palmetto Armory made pistols, muskets and swords for the State of South Carolina with machinery purchased from the Waters Armory of Millbury, Mass., until the contract was finished, and from 1861 until 1865 made cannon, minie rifle balls and 18 pdr. shells for the Confederacy. Probabilities are that flintlock muskets were altered to percussion at the armory during the Civil War, but no new arms were manufactured. The Palmetto Armory was burned by Sherman in February, 1865, but was rebuilt and later known as Palmetto Iron Works, or Shields' Foundry. Other Confederate armories scattered through the South made copies of the Model 1842 pistols during the war.

The firm was in existence until several years ago, and the building still standing, although in rather dilapidated condition, at the northeast corner of Lincoln and Laurel Streets.

U. S. PERCUSSION PISTOL, MODEL 1843, N. P. AMES, ARMY, NAVY
Illustrated—Fig. 2, Plate 6.

Caliber .54, taking a half ounce spherical ball. Six inch round smoothbore barrel without sights. Total length eleven-and-five-eighths inches. Weight 2 pounds. The barrel is lacquer browned; the lock and hammer are case-hardened. All mountings are brass. The flat butt has a rounded brass butt plate counter-sunk flush with the bottom. The three-quarter length walnut stock extends to the swivel ramrod. The lock is of the type called box-lock, the hammer being inside the lock-plate to facilitate handling when carried inside the sailors' belts.

The barrel of the specimen illustrated is marked "USN", "RP" and "P", and is dated 1844. The flat lock-plate is marked "N.P.AMES", "Springfield Mass." and "USN 1844".

The pistol was made under contract by N. P. Ames of Springfield, Mass., who also made the side-hammer Jenks carbines for the Navy. Pistols of this model were also made by Henry Deringer of Philadelphia, and were made both smoothbore and rifled. 2,000 are believed to have been made by Ames, and a lesser number by Deringer.

Though these pistols are called Model 1843, they are the first percussion pistols issued to the services, as they were issued before the 1842 model. The navy pistols are marked "U. S. N." Those made for army use were stamped "USR", for United States Mounted Rifles. The pattern pistols for this model were made at the Springfield Armory.

U. S. Percussion Pistol, MODEL 1843, DERINGER, Army, Navy

Illustration as for N. P. Ames above.

The pistol is identical in all respects with the Model

1843, N. P. Ames pistol described above, except for the marking on the lock-plate, which is stamped "US DER-INGER PHILADELA". Specimens are also known marked "USN" and dated behind the hammer, in addition. The barrels of some were marked "DERINGER PHILADELA RP", some only "RP", while some were unmarked.

The rifled pistols of this model are equipped with front and rear sights, and are rifled with seven grooves. The smoothbore pistols have a brass, blade front sight only.

The pistol was made by Henry Deringer, a rifle and pistol manufacturer of Philadelphia, Penn. Deringer was the son of Henry Deringer (Sr.), a Colonial gunsmith of German descent, maker of Kentucky rifles. Young Henry was born October 26th, 1786, at Easton, Pennsylvania, and as a youth was apprenticed to a firearm maker at Richmond, Virginia, where he made rifles and other fire-arms, until he settled in Philadelphia in 1806, and established an arms manufacturing plant of his own.

He is known to have made martial pistols and muskets in 1808, and later obtained the following contracts in addition to a contract of July 23, 1819, the details of which are not available:

April 23, 1821 — 2,000 rifles at $15.50 each.
Aug. 28, 1823 — 3,000 rifles at $14.50 each.

Lock-plates and rifles by Deringer dated 1841, indicate that he also had some later contracts.

The Deringer Armory though known for its pistols, was famous for the excellent rifles it produced, and later became well known for the small percussion pistols manu-factured by the firm, one of which was used by Booth to assassinate Lincoln.

The factory was located for many years on Front Street in Philadelphia. The story is that Deringer, a capable business man and shrewd trader, often traded squirrel rifles made at his factory for cargoes of lumber brought down the Delaware to Philadelphia. Since the traders came from all the settled parts of the country, Deringer arms were distributed over a wide area and created a further demand.

Deringer was content with the percussion system and stubbornly refused to manufacture breech-loaders which he despised. He died in 1868, and not many years after his death his factory went out of existence.

U. S. PERCUSSION PISTOL-CARBINE, MODEL 1855, SPRING-FIELD, ARMY.

Illustrated—Fig. 3, Plate 6.

Caliber .58. Twelve inch round barrel rifled with three broad grooves. Length of the pistol only, seventeen-and-three-quarter inches. Weight of pistol alone, 3 pounds, 13 ounces. All mountings are brass. The barrel has a low-blade front sight and a triple leaf rear sight graduated to 400 yards with a peep sight at 300. The full length walnut stock is oil finished and is reinforced at the forward end with a brass stock-tip. The ramrod button has a conical recess to seat the pointed bullet; the small end is threaded inside to take a wiper head or a bullet screw. The stock is equipped with a steel sling swivel on the band, and a pommel ring on the butt cap.

The pistol is furnished with a brass mounted, detach-able, walnut stock which may be readily attached to the

steel back-strap and tightened in place by a round nut.
When assembled with stock, the arm is twenty-eight-and-
one-quarter inches long and weighs about 5 pounds, 7
ounces. The stock also carries a sling swivel.

The lock is equipped with a Maynard priming magazine from which a coiled, ten inch strip of varnished
paper tape containing fifty spaced pellets or primers was
fed to the cone, one at a time, automatically, by the cocking of the hammer. The pistol would also function with
ordinary copper caps; but since the small cap was difficult
to handle mounted, the tape primer was considered a great
improvement.

The pistol used a three-groove conical bullet made in
two weights. The usual load was a 500 grain bullet with
a wedge-shaped hollow base which took a 60 grain charge
of black powder. A lighter 450 grain bullet with a hollow
base, shaped like a truncated cone used only a 40 grain
load.

In the specimen illustrated the barrel is marked "V",
"P" and the head of an eagle, the usual proofmarks of the
Springfield Armory, and is dated with the model year,
1855, on the tang. The flat bevelled-edge lock-plate is
marked "U. S. SPRINGFIELD" forward of the hammer,
and dated 1855 behind the hammer. A spread eagle is
stamped on the Maynard primer recess cover. Both the
pistol butt cap and the brass yoke of the shoulder stock
fastening device, are marked with number 12. The top
of the stock butt plate is marked "US".

Production records of the Springfield Armory show
a total of 4,021 pistol-carbines made; 2,710 in 1856 and

1,311 in 1857. Some experimental similar models were made at the Harper's Ferry Armory, but no record of production is available.

The pistol-carbine was adopted to supersede the .69 caliber musketoon for mounted troops in 1855, in which year the caliber of all long arms was changed from .69 to .58, and was issued to the cavalry and dragoon regiments of the regular army in 1856. The dragoons, whose tacti-cal employment was that of mounted infantry, found the arm satisfactory. It was light, powerful, accurate, and when dismounted into two parts, the stock and the pistol, was easily transported in saddle holsters. Since the dra-goons fought on foot, it was habitually used by them as a carbine, and they became adept in its use. The pistol-carbine however, was never popular with the cavalry. The point of strike differed when used as a pistol, and when used as a carbine with the steadying effect of the stock, and the men had but little confidence in the arm.

It is interesting to note that while the production of long arms at the Springfield and Harper's Ferry Armories, from their establishment until the beginning of the Civil War, totalled well over one million arms, less than ten thousand pistols, flintlock and percussion, had been made in both armories by 1860. The Ordnance Department charged with the procurement of arms, preferred in the main to buy or contract for its pistols with private manu-facturers, and so encourage the continuance, expansion and progress in the art of manufacture of these arms by private firms. For this reason government made, martial pistols are relatively scarce, and the earlier flintlocks are quite rare.

A small number of pistol-carbines very similar to the Springfield Model 1855 pistol-carbine described above was manufactured at the Harper's Ferry Armory. The year of manufacture is not certain but judging from the fact that the caliber of this Harper's Ferry pistol-carbine is .58, it was most likely made in 1853 or 1854, as a Board of Ordnance Officers began the study of small-arms calibers in 1853 and recommended the adoption of caliber .58 in 1855. No data is available as to the number made, but the issue was certainly small and probably experimental, possibly being the pattern followed later in the Springfield Model 1855.

These Harper's Ferry pistols have a twelve-inch round barrel, semi-octagonal at the breech and rifled with three grooves. Length of the pistol, eighteen inches; total length with stock, twenty-eight-and-one-quarter inches. Length of stock alone, eleven-and-one-half inches. The mountings are brass. Swivel ramrod. A steel knife-blade front sight is brazed on the barrel.

The Harper's Ferry pistol-carbine is dissimilar from the Springfield in that the lock-plate is not cut out for the Maynard primer magazine recess, and there is no rear sight. The barrel of the pistol is marked "P", "V", and an eagle head. The lock-plate is marked "PISTOL CAR-BINE" at the front, "HARPERS FERRY" in the center, and "U.S." vertically, reading to the rear, behind the hammer.

Chapter 3

U. S. MARTIAL SINGLE SHOT CARTRIDGE PISTOLS

THE DEVELOPMENT OF THE METALLIC CARTRIDGE—CARTRIDGE SINGLE SHOT PISTOLS

MODEL 1866—REMINGTON NAVY PISTOL.

MODEL 1867—REMINGTON NAVY PISTOL.

MODEL 1869—SPRINGFIELD ARMY PISTOL.

MODEL 1871—REMINGTON ARMY PISTOL.

UNITED STATES SINGLE SHOT CARTRIDGE PISTOLS

Though metallic and paper cartridges had been known for some hundreds of years prior to the invention of the percussion cap, they had merely served as powder and ball containers which made the handling of the charge more convenient and the loading more rapid.

The metallic cartridge was born of the desire to avoid the loading of arms from the muzzle, with its attendant impossibility of obtaining a tight fit of the bullet against the rifling, requisite for accuracy of fire. Muzzle loading also failed to utilize the entire effect of the propellant gasses, and necessarily resulted in loss of velocity and higher trajectories, due to the difficulty of making a bullet smaller than the bore take the rifling effectively, even the expanding, hollow base, conical bullet of the 'sixties.

The principle of breech-loading which permitted the use of a bullet larger than the bore and overcame the principal objections of the muzzle-loader, was used in our service as early as in 1817, when the government, after trials, ordered 100 breech-loading rifles invented by Capt. John Hall in 1811. Many other breech-loading arms, largely carbines, were invented between 1845 and 1860, but where the muzzle loader was objectionable because of the gas escape by the ball, these early breech-loaders were characterized by the escape of gas at the breech, and were not popular until the invention of the thin-walled metallic cartridge, which through expansion of the metal in the breech prevented gas leakage.

The first practical metallic cartridges were invented by Mr. Flobert of Paris about 1847, and were similar to the still muchly used caliber .22 BB rim fire cartridges of today. These early metallic cartridges were produced with the object of safety in handling, rapidity in loading, and protection from the elements, and were made in small calibers. It was thought that the heavy charge of a large caliber weapon required a proportionately thick casing, else the charge of a thin cartridge case would blow backwards; on the other hand, a thick cartridge case would have offered too much resistance to the blow of the hammer to ignite the charge.

The discovery that the thin-walled metallic cartridge had the properties of obturation, or sealing the breech against gas escape through the expansion of the casing, revolutionized the design of pistols and revolvers. This discovery was claimed by Dr. Maynard, inventor of the tape primer, who was also the patentee of a percussion carbine using a wide flanged brass cartridge with a hole in the center of the base, through which the flash of the ordinary percussion cap on a cone, was conducted to the charge in the cartridge.

In the eighteen-fifties, in addition to the cartridges of the Maynard type, of which a number were developed, some freak cartridges were brought out, such as tit-fire and cup primer, in the effort to circumvent the Rolin White patents controlled by Smith & Wesson for a bored-through cylinder.

The next development was the old style center fire cartridge with a concealed primer, such as was used in the early Smith & Wesson revolvers and Springfield breech-loading rifles. The next step was a combination of an outside primer and the metallic cartridge case of today. The story is that the idea of the removable center fire primer was developed from the successful attempts of the Sioux Indians to reload empties of the hidden primer type by boring out a hole in the base and inserting a percussion musket cap. The yarn may be true, but it is Colonel Hiram Berdan, a famous shot of Civil War days, arms inventor and patentee of the rifle bearing his name, who is given the credit for the "invention" of fixed ammunition of the present day, made possible by the discovery of the drawing process in which a disc of annealed brass was forced through a series of dies, forming a case with a base thicker than the walls; a base thick enough to contain the primer, and rigid enough to "back" it.

U. S. SINGLE SHOT NAVY PISTOL, MODEL 1866, REMINGTON

Illustrated—Fig. 1, Plate 7.

Caliber .50, rim fire. Eight-and-one-half inch round barrel rifled with three grooves and fitted with a round base, blade front sight and an open V-shaped notch rear sight cut into the top of the breech block. Total length thirteen-and-one-quarter inches. Weight 2 pounds, 4 ounces. The barrel was blued; the receiver, trigger guard, trigger, hammer and breech-block were case-hardened.

In the specimen illustrated the left side of the receiver is stamped "REMINGTONS ILION N.Y. U.S.A. PAT. MAY 3d NOV. 15th 1864, APRIL 17, 1866". On the right is stamped "P FCW". An anchor is stamped on the barrel, and inspector's initials "HW" in script are stamped in a medallion on the left side of the walnut grip. The pistol fired a rim fire cartridge with a conical bullet.

This is the first model of the Remington single shot pistols and has a *sheath trigger*. The design and patent were by Joseph Rider, a celebrated firearms inventor. The keynote of the Remington-Rider system lies in the utilization of the hammer in its down position to both lock and support the breech-block.

After the first year the sheath-trigger of this model was abandoned in favor of the conventional trigger and trigger guard of the Model 1867. The sheath-trigger was found unsatisfactory because of the danger of accidental discharge when the pistol was thrust into a holster, cocked.

Only five hundred of these pistols are reputed to have been made and they are quite rare.

The Remington Arms Company was founded by Eliphalet Remington, Jr., and his father in 1816, when the young Remington, assistant in his father's blacksmith shop, turned to the manufacture of firearms. By 1856, when Remington took his three sons into the business, the firm had fulfilled a number of government contracts. The Remingtons furnished the Navy Department with Jenks patent side-hammer carbines; took over and completed by 1850 a contract for 5,000 Model 1841 rifles in 1845 from John Griffiths of Cincinnati, and later furnished 7,500 additional like arms. In 1857-58 the firm modified 5,000 muskets for the army by attaching Remington primer locks.

During the Civil War, Remingtons furnished the government with 10,000 Model 1855, sabre bayonet rifles, and 39,000 Model 1863 rifle muskets. Remington also furnished 125,314 Remington percussion revolvers, and 2,814 Beals revolvers. In 1865 the company was incorporated and secured the services of Joseph Rider, the famous arms inventor, and enjoyed a period of prosperity until 1866, when it failed and was reorganized as the Remington Arms Company. The control of the business passed from the Remington family to Hartley & Graham of New York. The company was merged in 1902 with the Union Metallic Cartridge Company and became known as Remington-UMC. Later the name was changed again to Remington Arms Company, Inc.

U. S. SINGLE SHOT NAVY PISTOL, MODEL 1867, REMINGTON
Illustrated—Fig. 2, Plate 7.

Caliber .50, center fire. Seven inch round barrel rifled with three grooves and fitted with a round base blade front sight. An open V-shaped rear sight is cut into the top of the breech block. Total length eleven-and-three-quarters inches. Weight 2 pounds. The barrel was blued; the receiver, trigger guard, trigger, hammer and breech-lock were case-hardened.

In the specimen illustrated the barrel is marked "I HE", and is stamped with an anchor. The left side of the receiver is marked "REMINGTONS ILION N.Y. U.S.A. PAT. MAY 3d NOV 15th 1864. APRIL 17th 1866". On the right side of the receiver are letters "PFCW". The pistol number 2061 is stamped on the barrel and inside the grip frame. The left side of the walnut grip frame is marked with inspector's initials "FCH" in script in a medallion. The pistol used a center fire cartridge with a blunt nose, conical bullet weighing 300 grains, propelled by 25 grains of black powder.

This is the second of the Remington Navy pistols and differs from the first model in having a regular oval trigger guard and a shorter barrel, as well as being made for a center fire cartridge. A total of 6,500 of these pistols were furnished to the Navy: 5,000 under contract of Nov. 14th, 1866, which was extended for an additional 1,500.

U. S. SINGLE SHOT ARMY PISTOL, MODEL 1869, SPRINGFIELD
Illustrated—Fig. 3, Plate 7.

Caliber .50, center fire. Nine inch round barrel rifled with three grooves. Total length eighteen-and-one-quarter inches. Weight 4 pounds, 8 ounces. The barrel is fitted with a large blade front sight; a V-notch rear sight is cut in the receiver. The barrel is held to the stock by a musket type band, spring-fastened in front. The stock extends to within three-and-one-half inches of the muzzle. Brass butt cap and back-strap. Steel trigger guard and extensions. The barrel and the trigger guard were blued, the receiver, breech-block and lock-plate were case-hardened.

The mechanism of the pistol is that of the Allin system, Springfield Model 1868 rifle. The lock-plate and the lock mechanism are those of a Civil War rifle musket.

The receiver of the specimen illustrated is marked "DEC 1869", the breech-block is stamped "1869", an eagle and "U.S." The flat musket lock-plate is marked with an eagle, "U.S." and "SPRINGFIELD" in front of the hammer, and "1863" in rear.

This huge, heavy and clumsy arm using a 50-50 Government C.F. rifle cartridge, was made experimentally at the Springfield Armory from musket and rifle parts.

In the days of the revolver this unwieldly arm with its tremendous charge was hardly a success, and the manufacture never went beyond the experimental stage.

U. S. Single Shot Army Pistol, MODEL 1871, REMINGTON
Illustrated—Fig. 4, Plate 7.

Caliber .50, center fire. Eight inch round barrel rifled with three grooves. The barrel carries a large knife blade front sight. An open V-shaped notch is cut into the top of the breech-block. Total length twelve inches. Weight 2 pounds. The barrel and trigger are blued; the frame is case-hardened in mottled colors. The hammer and breech-block are burnished bright.

The specimen illustrated is marked on the left side of the receiver "REMINGTONS ILION N.Y.U.S.A.PAT. MAY 3d NOV. 15th 1864. APRIL 17th 1866" and letters "PS". Number 4936 is stamped on the grip frame and inside the walnut grip. The left side of the grip bears inspector's initials "CRS" in script in a medallion. 5,000 of these pistols were furnished to the army in 1870.

The pistol differs from the Navy Model 1867 in length of barrel, better front sight, and fine balance, but the principal difference is in the grip, which is of a rounded fish-tail shape, made to fit the hand and curving back to fit the knuckle of the thumb. This improvement was made on the recommendation of an Ordnance Department Board.

As a military arm it was hardly a success. In the era of the revolver this single shot pistol was an anachronism of the horse pistol days. But if a failure from the military standpoint, and soon supplanted by the revolver; because of its splendid balance and fine grip, it was very popular as a target pistol. Though no longer made, it is still sought for by "gun-cranks" for target shooting when modified by rebarreling to smaller calibers.

SECONDARY MARTIAL PISTOLS

NOTES ON SECONDARY MARTIAL PERIODS —

SECONDARY PISTOLS

ANSTAT
BIERLY
BIRD
BOOTH
CALDERWOOD
CHERINGTON
COUTTY
DERINGER
DERR
DREPPERD
ELGIN
EVANS
FRENCH
GRUBB
GUEST
HALL
HENRY, J.
HENRY, J. J.
KLINE
KUNTZ
LINDSAY
MARSTON
McK. BROTHERS
MEACHAM & POND
MILES
MILLS
MOLL
PERKIN
PERRY
POND
RICHMOND
ROGERS
RUPP
SHARPS
SHULER
SWEITZER
VIRGINIA
WALSH
WATERS

UNITED STATES MARTIAL SECONDARY PISTOLS

The classification of the regular issue, or primary United States martial pistols described in the foregoing pages includes those arms made at the government armories, or under contract at the private armories such as those of Simeon North, Asa Waters, Robert Johnson, Henry Aston and others, whose arms were used to equip the regular army, the navy and the militia of the several states. In addition to these government recognized armories whose output was marked "US", a number of other private arms manufacturers, as well as some of the official contractors, made martial pistols for sale to the states direct for militia use; to privateers; to citizen military organizations, and to individual officers for personal arms. Other makers made pistols for sale to the government or with expectations of such disposal, and many experimental and freak arms were made with the hope of government adoption for use in the armed services. These single barrel martial type arms, to follow Mr. Sawyer's classification, are described in this chapter under the heading of United States Secondary Pistols.

Though the Secondary pistols described and illustrated in this chapter do not cover the products of all the many of our early martial pistol makers, they are representative of the better known or famous makes, and include all those which were available for personal examination in a number of large and well known collections.

Any comparison of these arms must allow for the fact that the vast majority of these arms were the hand made, limited quantity products of individual craftsmen, and often considerable variation exists in barrel lengths, marking and design, even in the same models of identical makers.

In this chapter are also included a few pistols, mostly of the Model 1808 type, such as Deringer, French, Henry, Shuler, which though marked "US", were contracted for or purchased by the government in small quantities for militia use during the War of 1812. But few of these pistols were made, as is indicated by their present day rarity, and since they were not obtained in sufficient numbers for general issue, these arms are classified in this volume as Secondary pistols.

Finally, a number of pistols are included in which the borderline between the duelling and martial types has been drawn rather fine, for in those days many an arm could be made to serve a dual purpose.

ANSTAT FLINTLOCK PISTOL
Illustrated—Fig. 1, Plate 8.

Caliber .54, taking a half-ounce spherical ball. Eight-and-five-eighths inch round, smoothbore barrel, octagonal at the rear half. Total length thirteen-and-one-half inches. Brass mountings. The barrel and the thimbles are pin fastened. The full length, curly maple stock ends flush with the muzzle. The barrel carries a brass front sight. There is no rear sight or back-strap. The curved-in butt has a shallow butt cap. The brass trigger guard forks at the rear into a short branch. Flat goose-neck hammer. Bevelled iron pan with fence. Hickory ramrod.

The barrel of the pistol illustrated is marked "AN-STAT" in engraved block letters. The flat lock-plate has a double groove at the rear and is unmarked.

Anstat is believed to be Peter Angstadt of Lancaster, Pennsylvania.

BIELRY & CO., FLINTLOCK PISTOL
Illustrated—Fig. 2, Plate 8.

Caliber .54, taking a half-ounce spherical ball. Ten inch round, smoothbore barrel. Total length sixteen inches. Weight 2 pounds, 3 ounces. Silver mountings. The barrel is key fastened to the stock and has a rib extension underneath, from the end of the stock to the muzzle. The rib carries an iron thimble. On the barrel is mounted a small, silver front sight; a large open rear sight is in the tang. The walnut half-stock is tipped with horn at the forward end. The oval grip is checked, and ends in a spreading butt with a flat bottom. There is no butt cap or back-strap. The silver trigger guard forks at the rear to complete the oval. Flat, bevelled-edge, engraved goose-neck hammer; iron pan with fence. The frizzen spring is equipped with a roller. Hickory, brass-tipped ramrod.

In the specimen illustrated the flat, bevelled-edge lock-plate is marked "BIELRY & CO." in two lines, between the hammer and the frizzen spring, and is engraved behind the hammer. The barrel is unmarked.

C. BIRD & CO., Flintlock Pistol
Illustrated—Fig. 3, Plate 8.

Caliber .58. Twelve inch semi-octagonal smoothbore barrel carrying a brass blade front sight, and an open V-notch rear sight. Total length seventeen inches. Weight 2 pounds, 5 ounces. Iron mountings. The barrel and the two thimbles are pin fastened. The full length walnut stock extends to within one-quarter of an inch of the muzzle, and is rounded at the butt. There is no butt cap or back-strap. The iron trigger guard forks at the rear to complete the oval. Goose-neck hammer, iron pan with fence. The frizzen spring is equipped with roller. Hickory ramrod.

The barrel is unmarked. The flat lock-plate is marked "C. BIRD & CO. PHILADa WARRANTED", in three lines between the hammer and the frizzen spring.

Bird brothers, Philadelphia gunsmiths were active from about 1812 to 1820.

C. BIRD & CO., Flintlock Pistol
Illustrated—Fig. 4, Plate 8.

Caliber .58. Twelve inch semi-octagonal, smoothbore barrel carrying a brass blade front sight. Total length seventeen inches. Weight 2 pounds, 5 ounces. Iron mountings. The barrel and the two thimbels are pin fastened. The full length walnut stock extends to the muzzle. The oval butt swells to a rounded rim at the flat bottom. There is no butt cap or back-strap. The iron trigger guard curves back and down the butt without forking. Goose-neck hammer; iron pan with fence; roller equipped frizzen spring.

The barrel is unmarked. The lock-plate is marked "C. BIRD & CO. PHILADa WARRANTED" in three lines between the hammer and the frizzen spring.

The pistol dates to about 1820.

BOOTH FLINTLOCK PISTOL

Illustrated—Fig. 5, Plate 8.

Caliber .58. Eight inch brass, round, tapered, smooth-bore barrel without sights. Total length thirteen-and-one-half inches. Weight 1 pound, 13 ounces. Brass mountings. The barrel and thimbles are pin fastened. The full length walnut stock extends to within one-eighth of an inch of the muzzle. The butt is rounded without back-strap or butt cap. Goose-neck hammer. The frizzen curls at the heel. The iron pan is rounded at the bottom and has a fence. The brass trigger guard forks at the rear to complete the oval. Hickory swell tipped ramrod.

The barrel is marked "PHILADELPHIA" and is not proofmarked. The flat lock-plate is sunk flush with the wood, and is marked "BOOTH" between the hammer and the frizzen spring. The pistol probably dates to about 1800.

Booth is listed in the Philadelphia City Directory from 1798 until 1816. In addition to the manufacture of pistols of martial type, he is reputed to have specialized in the making, furnishing and rental of pistols for duelling purposes, when duelling was the vogue for settlement of "affairs of honor", though the young bloods of that era usually possessed their own pistols of that type.

CALDERWOOD Flintlock Pistol, MODEL 1808 Type

Caliber .54, taking a half-ounce spherical ball. Ten inch round smoothbore barrel. Total length sixteen inches. Brass mountings. The barrel and thimbles are pin fastened. The full length stock extends to within one-eighth of an inch of the muzzle. The sloping, slightly fish-tailed butt ends in a brass butt cap with rounded side extensions. There is no back-strap. The brass trigger guard forks at the rear to complete the oval. Brass blade front sight; there is no rear sight. Goose-neck hammer. The priming pan is of peculiar construction, being of brass and attached to the barrel. Hickory ramrod.

In the specimen described the flat, bevelled-edge lock-plate is marked "CALDERWOOD PHILA" between the hammer and the frizzen spring, and "US", "1808" vertically behind the hammer. The barrel bears the proof-mark "P" and an eagle head.

To judge by the marking on the lock-plate the pistol was made on government contract of 1808 for militia use. It closely resembles in general appearance the Model 1808 North pistol which was made in accordance with a pattern submitted by the government. The rarity of this arm would indicate that the Calderwood pistol contract or purchase was for a very small quantity of arms.

William Calderwood, Gunsmith, is listed in the Phila-delphia City Directories as residing on Germantown Road, in the issues 1813 to 1819 inclusive, (except 1812 and 1815 when no directories were published).

T. P. CHERINGTON FLINTLOCK PISTOL
Illustrated—Fig. 1, Plate 9.

Caliber .45. Twelve-and-one-quarter inch octagonal, smoothbore barrel. Total length seventeen-and-five-eighths inches. Weight 3 pounds, 1 ounce. The barrel and the brass thimbles are pin fastened. The full length walnut stock ends in a brass end-cap close to the muzzle. A brass blade front sight is mounted on the barrel; there is no rear sight. The butt is bird-head shaped, without butt cap or back-strap. The trigger guard is iron, and forks at the rear to complete the oval. Flat, bevelled-edge, goose-neck hammer; brass pan with fence. Hickory brass tipped ram-rod.

The flat lock-plate of the pistol illustrated is marked "T. P. CHERINGTON" between the hammer and the frizzen. The barrel is likewise marked "T. P. CHERING-TON".

The pistol was made by Thomas P. Cherington, Penn-sylvania rifle maker of Cattawissa, Pa.

COUTTY FLINTLOCK PISTOL
Illustrated—Fig. 2, Plate 9.

Caliber .58. Seven - and - three - quarters inch *brass,* tapered, smoothbore barrel without sights. Total length thirteen-and-one-half inches. Weight 1 pound, 11 ounces. Brass mountings. The barrel and the two thimbles are pin fastened. The full length walnut stock extends almost to the muzzle. The butt is rounded at the heel and has a brass butt cap, curving upward at the rear. Goose-neck hammer. The iron pan has a fence at the rear. The brass trigger guard forks at the rear to complete the oval, the rear extension reinforcing the grip. Hickory swell tipped ramrod.

The barrel is marked "PHILADELPHIA", "P" in a circle, "PV" and "V". The flat lock-plate is sunk flush with the stock, and is marked "COUTTY" between the hammer and the frizzen.

Samuel Coutty, listed in the Philadephia City Directory as residing in that city from 1785 to 1794, made arms for private sale and worked on public arms for the Commonwealth of Pennsylvania.

H. DERINGER, FLINTLOCK PISTOL, MODEL 1808 TYPE
Illustrated—Fig. 3, Plate 9.

Caliber .52. Ten inch round smoothbore barrel. Total length sixteen-and-one-half inches. Weight 2 pounds, 9-1/2 ounces. Brass mountings. The barrel, trigger-guard and the two thimbles are pin fastened. The full length walnut stock extends to within one-quarter of an inch of the muzzle. On the barrel is mounted a pyramidal-shaped, rounded top front sight; there is no rear sight. The rounded butt is reinforced by a brass butt cap with short rounded extensions reaching into the stock on each side. The brass trigger guard forks at the rear into a graceful curl to complete the oval. Flat, bevelled, double-necked hammer. Horizontal iron pan with fence. Hickory, swell-tipped ramrod.

In the specimen described, the flat, bevelled-edge lock-plate ends in a point and is marked "H. DERINGER PHILA", between the hammer and the frizzen. The barrel is proofmarked "P" in a circle.

The pistol was made by Henry Deringer and closely resembles the North Model 1808 Navy pistol. Very likely it was made under contract of 1808, after the same patterns and for militia use. (See Deringer, Henry.)

H. DERINGER FLINTLOCK PISTOL, MODEL 1826 TYPE

A very few Deringer flintlock pistols are known closely resembling the Model 1826 S. North pistols, except that the spring-fastened band is not musket band shaped but is of equal width throughout; the stock ends flush with the end of the band, and the grip is straighter and more bulbous at the butt than the 1826 North. The total length of the arm is fifteen inches.

Like the North Model 1826, the caliber is .54, the furniture is iron and the pistol is equipped with a swivel ramrod, a double-necked hammer and a tilted brass pan without fence.

The lock-plate is flat with bevelled edges at the front and the surface is slightly rounded at the rear. The lock-plate is marked "US", "H. DERINGER" and "PHILA-DA", in three lines, between the hammer and the frizzen spring. The date "1826" is stamped vertically behind the hammer.

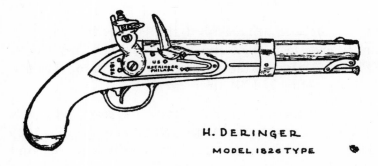

H. DERINGER

MODEL 1826 TYPE

Probabilities are that a few pistols of this pattern were made up and submitted to the government in hope of a contract that did not materialize.

JOHN DERR Percussion Pistol
Illustrated—Fig. 4, Plate 9.

Caliber .54, taking a half-ounce spherical ball. Ten-and-three-eighths inch round, smoothbore barrel, octagonal at the rear. Total length sixteen inches. Brass mountings. The barrel is key fastened to the stock; the thimbles are pinned. The full length maple stock ends in a brass end-cap flush with the muzzle. The bird-head shaped, deep butt has a shallow butt cap. There is no back-strap. The barrel carries a long, brass blade front sight and an open, V-notch, iron rear sight. The brass trigger guard forks at the rear. The trigger curls at the bottom. Cylindrical side lug. Hickory ramrod.

The barrel of the specimen illustrated is marked

"JOHN DERR WARRANTED". The flat lock-plate is unmarked.

The pistol was made by John Derr, a Lancaster, Penna., gunsmith.

DREPPERD PERCUSSION PISTOL
Illustrated—Fig. 5, Plate 9.

Caliber .40. Eight-and-seven-eighths inch octagonal, *brass*, smoothbore barrel. Total length fourteen-and-one-half inches. Brass mountings. Key fastened barrel and pin fastened thimbles. The full length maple stock ends in a brass end-cap within one-sixteenth of an inch of the muzzle. A brass, blade front sight is mounted on the barrel; the wide rear sight is placed on the top of the tang. The rounded butt ends in a plain brass cap. There is no backstrap. The brass trigger guard forks midway of the loop at the rear, the graceful forward curl completing the oval. The lock-plate and hammer are engraved all over. The cylindrical percussion cap lug is slotted to facilitate removal.

The specimen described is marked on the lock-plate "DREPPERD LANCASTER" in front of the hammer. The barrel is unmarked.

The Lancaster, Penna., City Directory for 1857 lists two Drepperds, John and Andrew, among the Lancaster gunsmiths.

ELGIN Cutlass-Pistol (C. B. ALLEN Make)
Illustrated—Fig. 1, Plate 10.

Caliber .54, single shot. Five inch octagonal smooth-bore barrel with an iron, blade front sight. There is no rear sight. Total length fifteen-and-three-quarters inches. Weight 2 pounds, 7 ounces. Side hammer, iron frame and back-strap, walnut grips.

A knife blade eleven-and-one-half inches long and two-and-one-sixteenth inches wide is fastened under the barrel and in front of the trigger guard. At the rear, the iron trigger guard extends in a large loop to the butt to form a hilt or a hand-guard. A steel, brass tipped ramrod was carried in a leather sheath with which the pistol was equipped. The pistol was loaded with loose powder and ball.

The specimen illustrated is marked on the barrel "EL-GIN'S PATENT PM CBA 1837", and on the left side of the frame "C. B. ALLEN SPRINGFIELD MASS." Number 149 is stamped on the left side of the barrel, the left side of the frame, and the left side of the blade.

This unique arm was the invention of George Elgin of New York, who had hoped to have it adopted as a naval boarding arm to replace both the cutlass and the pistol in the naval service. The manufacturer of this model was C. B. Allen, a gunsmith of Springfield, Mass., 1836-41. Probabilities are that a number of sizes were made. Mr. Sawyer describes one with a thirteen-and-one-quarter inch blade, seventeen-and-one-half inches over all.

ELGIN Cutlass-Pistol (MORRILL, MOSMAN & BLAIR Make)

Illustrated—Fig. 2, Plate 10.

Caliber .38, percussion, single shot. Four inch octagonal barrel rifled with five grooves, and carrying a small brass blade front sight and an open V-notch rear sight. Total length fourteen-and-one-half inches. Weight 1 pound, 10 ounces. Side hammer, iron frame, maple stock.

A knife blade nine-and-one-half inches long and one-and-one-half inches wide is mounted under the barrel. The rear of this blade forms the trigger guard. There is no ramrod. The arm is equipped with a leather scabbard with belt hook. The pistol was loaded with loose powder and ball.

The specimen illustrated is marked in etched words; on the right side of the blade "ELGIN'S PATENT", and on the left side "MORRILL MOSMAN & BLAIR AMHERST MASS."

This is another model of the Elgin patent cutlass-pistol. Other variations are known differing in minor details and size, including a Bowie-Knife Pistol with a three inch barrel and a seven inch blade.

The specimen illustrated was manufactured by a cutlery and machinery manufacturing concern organized April 1, 1836, by Henry A. Morrill, Silas Mosman, Jr., and Charles Blair. According to an article in the Boston Courier, issue of April 30, 1837, . . . "the manufacturer has a contract for one thousand . . . Bowie-Knife Pistols . . . for a Georgia man who is the patentee. They are made in three sizes." . . .

The business panic of 1837 affected the enterprise and the partnership was dissolved in July, 1838, the business being carried on by Silas Mosman and Charles Blair until February, 1839, when the firm failed, and the machinery and the effects were sold at an assignee's sale.

In this connection an advertisement published in the Hampshire Gazette under date of March 8, 1837, may be of interest: —

> "WANTED — Six or eight filers, who can do first-rate work, and feel smart enough to do a day's work in ten hours, without raising higher pressure of steam than cold water will make, and can leave their long yarns until their day's work is done. Such will find good encouragement by applying immediately to
>
> MORRILL, MOSMAN & BLAIR."

EVANS Flintlock Pistol, FRENCH MODEL 1805 Type

Illustrated—Fig. 1, Plate 11.

Caliber .69, taking an ounce spherical ball. Eight-and-seven-eighths inch round smoothbore barrel without sights. Total length fifteen inches. Weight 2 pounds, 13 ounces. Brass mountings. The barrel is held by a brass band fastened by a spring at the rear. The stock ends at a receding angle flush with the band, which juts forward at the bottom. The rounded butt is reinforced by an iron back-strap extending from the tang to the plain brass butt cap. Oval brass trigger guard; rounded face, double-necked hammer; brass pan without fence. Steel button-head ramrod.

The barrel of the specimen illustrated is marked with the proofmark "P" on the left side of the breech. The underside of the barrel is stamped "PM 1814". The flat bevelled-edge lock-plate is marked "EVANS" between the hammer and the frizzen spring.

The arm was designed after the French army pistol of the 13th Year of the Republic (1805). Probabilities are that it was made by the firm of O. & E. Evans of Evansburg, which on Oct. 25, 1808, obtained a U. S. contract of five years' duration for 4,000 stands of arms, of which 1,960 are known to have been completed and delivered by Oct. 7, 1812.* Aug. 14, 1815, O. & E. Evans obtained an additional contract of one year's duration, probably to close the previous contract account. At this time the firm was managed by Edward Evans, Owen having died in 1812.

Edward and Owen Evans had lived at Evansburg, Pa., a village about five miles north of Valley Forge. Owen's son, William L. Evans, later made Model 1821 muskets and Model 1826 pistols at the Valley Forge. (See W. L. Evans F/L pistol Model 1826.)

T. FRENCH Flintlock Pistol, MODEL 1808 Type

Illustrated—Fig. 2, Plate 11.

Caliber .64, taking an ounce ball. Ten-and-five-eighths inch round, smoothbore barrel. Total length sixteen-and-three-quarters inches. Brass mountings. The barrel and the two thimbles are pin fastened. The full length walnut stock extends to within three-eighths of an inch of the muzzle. The rounded butt ends in a brass butt cap, and is reinforced by the long, iron barrel tang which forms the

*Paper 116, American State Papers, Class V, Military Affairs.

back-strap. There are no sights. The brass trigger guard forks at the rear to complete the oval. Bevelled-edge, goose-neck hammer. Iron pan with fence.

The lock-plate of the specimen illustrated is marked with an eagle, "US" and "T. FRENCH" between the hammer and the frizzen spring, and "CANTON" vertically in rear of the hammer, in front of two grooves. The barrel is marked "P.M." "PC" in a rectangle and "1814".

Thomas French of Canton, Mass., (1778-1825) maker of this contract pistol for militia issue, on October 20, 1808, in partnership with Blake and Adam Kinsley, obtained a contract for 4,000 stands of arms. Of these 2,175 muskets are recorded to have been delivered by French, Blake & Kinsley by Oct. 7, 1812.

T. GRUBB Flintlock Pistol

Illustrated—Fig. 3, Plate 11.

Caliber .44. Eight-and-three-quarters inch round, *brass,* smoothbore barrel, octagonal for the rear half. The barrel tang is of iron. Total length fourteen-and-three-eighths inches. Silver mountings. The barrel is key fastened; the thimbles are pinned. The horn tipped, walnut stock extends to within one-eighth of an inch of the muzzle. A silver front sight is mounted on the barrel. There is no rear sight. The curved-in, bird-head shaped butt has a silver butt cap and is ornamented with silver inserts. The trigger guard is of silver, reinforced at the front, and forking at the rear into a curl. Flat goose-neck hammer with curled tip; iron pan with fence; curled frizzen end. Hickory ramrod with horn tip.

The barrel is marked "T. GRUBB". The flat un-marked lock-plate has two vertical grooves at the rear, and was probably imported.

The handsome arm illustrated, was made by T. Grubb, a Philadelphia gunsmith of the eighteen-thirties.

I. GUEST Flintlock Pistol, MODEL 1808 Type

Illustrated—Fig. 4, Plate 11.

Caliber .54, taking a half ounce spherical ball. Ten-and-one-quarter inch round, smoothbore barrel. Total length sixteen inches. Brass mountings. The barrel and the two thimbles are pin fastened. The full length walnut stock ends one-sixteenth of an inch from the muzzle. Brass blade front sight. There is no rear sight or back-strap. The butt ends in a brass butt cap with one inch long side extensions. The trigger guard forks at the rear to complete the oval. Flat, bevelled-edge, double-necked hammer. Iron pan with fence. Hickory swell-tipped ramrod.

The barrel is marked with an eagle head and "P" in an oval, and "I. GUEST" in script. The flat bevelled lock-plate is marked "DREPERT" between the hammer and frizzen spring and "US" behind the hammer.

I. Guest worked at the Warwick Iron Works, which cast cannon during the Revolutionary War. Drepert, the lock maker of this pistol is believed to be Henry Drepert, also spelled Dreppert, Drippard and Drepperd, a gunsmith of Lancaster, Penn., of the flintlock period.

HALL BREECH-LOADING FLINTLOCK PISTOL. (Bronze Barrel and Breech.)

Illustrated—Figs. 1 & 2, Plate 12.

Caliber .50, single shot. Five-and-nine-sixteenths inch *bronze,* octagonal, smoothbore barrel. Total length fourteen-and-five-eighths inches. Weight 2 pounds, 13 ounces. The barrel and the trigger guard are pin fastened. The full length walnut stock is capped with a brass end-band and extends to within one-quarter of an inch of the muzzle;—the rounded butt is checked. There is no back-strap or butt cap. The oval brass trigger guard is engraved with a simple design on the forward extension; — the rear of the guard loop forks to complete the oval. The pan forms a part of the bronze breech-block. Iron side-straps form the breech-block frame. Steel hammer and frizzen. Bronze latch frame. A steel blade front sight is wedge-mounted on the barrel and is offset to the left to clear the hammer and pan. There is no rear sight. The trigger is offset to the right.

The specimen illustrated and described is marked only "35" on the right side of the breech-block.

The pistol is similar in design to the Hall breech-loading rifles Model 1817, invented and patented by John H. Hall, and produced at the Harpers Ferry and the S. North Armories.

By pressing rearward and upward on a spur projecting in front of the trigger guard, the bronze breech can be raised and the chamber loaded with powder and ball, or prepared paper cartridge. The mechanism of this pistol differs from that of the rifle, in that the barrel and breech-block are bronze, and the frizzen spring is flat and lies along the top of the breech in front of the pan.

This pistol was made by Captain John H. Hall of Yarmouth, Maine, who in 1811 was granted a patent for a breech-loading firelock (flintlock). Between 1811 and 1816 Hall made a limited number of sporting arms and pistols embodying his system. Doubtless this pistol with its bronze barrel and breech was one of Hall's earliest products, and quite likely was hand made some time between 1811 and 1814.

About 1812 Captain Hall adopted his system to the heavier charge of martial long arms, and for a time vainly attempted to have them accepted by the services. Finally in January, 1817, after tests of 1813 and 1816, Hall was given a contract for one hundred rifles for service trials and tests. As a result of favorable reports on his arms, the rifle was officially adopted, and after another period of two years spent at the Harpers Ferry Armory perfecting the mechanism, Captain Hall in March, 1819, received a contract for one thousand breech-loading rifles bearing his name. In order to insure quantity production and proper construction, Hall entered government employ as assistant armorer at the Harpers Ferry Armory to supervise the manufacture of his arms, at a salary of sixty dollars per month and royalty of one dollar per rifle. In this connection it is interesting to note that in the production of these arms Hall followed in the footsteps of Simeon North* and designed and constructed a number of machines used in the manufacture of his rifles in order to insure the interchangeability of parts and facility of manufacture. This

*See S. NORTH Flintlock Pistol Model 1813.

was the first instance of practical standardization of parts in a government arms plant.

HALL BREECH-LOADING FLINTLOCK PISTOL (Iron Barrel and Breech)

Illustrated—Fig. 3, Plate 12.

Caliber .50, single shot. Seven-and-one-eighth inch octagonal iron barrel rifled with eight grooves. Total length sixteen-and-one-quarter inches. Weight 2 pounds, 15 ounces. The barrel is key fastened to the stock by two keys. The full length walnut stock extends to within one-sixteenth of an inch of the muzzle, and is reinforced near the end by a silver band. The checked butt is oval shaped and flat at the bottom and has a silver butt plate set flush. The silver trigger guard forks at the rear to complete the oval. The spur plate is also silver. Steel hammer and frizzen. The iron pan is integral with the iron breech-block. A brass blade front sight is offset to the left to compensate for the centrally located hammer and pan. There is no rear sight.

The pistol illustrated is marked on the top of the breech-block "JOHN H. HALL PATENT". The right side of the block is marked "RB 40".

Pressing rearward and upward on the locking catch or spur, in front of the trigger guard opens the breech upward, exposing the chamber for loading. The mechanism is similar to that of the Hall rifle, and has a rifle type, rider frizzen spring.

The arm described, one of a matched pair, was probably made about 1815, shortly before Hall's entry into government employment at the Harpers Ferry Armory to superintend the manufacture of his breech-loading rifles.

J. HENRY (Phila) Flintlock Pistol, MODEL 1808 Type

Illustrated—Fig. 1, **Plate 13.**

Caliber .54, taking a half ounce spherical ball. Ten inch round smoothbore barrel. Total length sixteen inches. Weight 2 pounds, 9 ounces. Brass mountings. The barrel, trigger and the brass thimbles are pin fastened. The full length stock extends to within one-quarter of an inch of the muzzle. A knife-blade brass front sight is mounted on the barrel. There is no rear sight. The tapering rounded butt ends in a brass butt cap with short, rounded side extensions. The brass trigger guard forks at the rear to complete the oval. The hammer is flat, bevelled, double-necked. Iron pan with fence. Hickory swell tipped ramrod.

The barrel is marked "J. HENRY PHILA" and is proofmarked with an eagle head and "P" in the same oval. The lock-plate is marked "U.S." between the hammer and the frizzen spring, and "J. HENRY PHILA" behind the hammer.

To judge by the marking on the lock-plate the pistol was made on government contract of 1808 for militia use. It closely resembles in general appearance the Model 1808 North pistol which was made in accordance with a pattern submitted by the government.

J. Henry pistols of this model are also known with barrels marked "P", and lock-plate stamped only "J. HENRY PHILA". Others were made with rounded butts and without butt caps.

J. Henry, maker of this pistol was probably Joseph Henry, a Philadelphia gunsmith, 1811-24. Joseph Henry was associated with John Joseph Henry, a relative, in the production and repair of public arms for the Committee of Defence of Philadelphia, during the War of 1812.

J. HENRY Flintlock Pistol

Illustrated—Fig. 2, Plate 13.

Caliber .62. Ten inch octagon, smoothbore barrel. Total length fifteen-and-three-quarters inches. Brass mountings. The barrel is key fastened; the thimbles are pinned. The full length walnut stock extends to within one-eighth inch of the muzzle. A brass, blade front sight is mounted on the barrel. There is no rear sight. The swelled, rounded butt ends in a brass butt cap with short, rounded side extensions. There is no back-strap. The plain, brass trigger guard protects a straight trigger. Flat, bevelled-edge, goose-neck hammer. Iron pan with fence. The end of the frizzen ends in an ornamental curve. Swelled hickory ramrod.

The flat, bevelled lock-plate is marked "J. HENRY" between the hammer and the frizzen spring. There are two vertical cuts behind the hammer, seven-sixteenths of an inch from the end of the plate. The barrel is unmarked.

The "Joseph Henry gun factory" is listed in the Philadelphia City Directory from 1811 to 1824, at the corner of N. 3rd and Noble Streets.

J. J. HENRY (Boulton) FLINTLOCK PISTOL

Illustrated—Fig. 3, Plate 13.

Caliber .58. Eight-and-seven-eighths inch round, smoothbore barrel, octagonal at the rear. Total length fourteen-and-one-half inches. Weight 2 pounds, 3 ounces. Brass mountings. The barrel and the two thimbles are pin fastened. The full length stock extends to the muzzle. Brass blade front sight; there is no rear sight. The rounded butt ends in a brass butt cap with short, rounded side extensions. There is no back-strap. The brass trigger guard forks at the rear to complete the oval. The goosenecked hammer is flat, bevel-edged. Iron pan with fence. The frizzen spring is equipped with a roller. Both the lock-plate and the hammer are engraved. Hickory ramrod.

The flat, bevelled-edge lock-plate is marked "J. J. HENRY BOULTON" between the hammer and the frizzen spring. The barrel is unmarked.

Similar Henry pistols made for militia use, are known, varying in minor details and marking.

The Henry firm of gunsmiths was founded by William Henry I of Chester County, Pa.,who upon finishing his apprenticeship to Mathew Roeser, gunsmith of Lancaster, Pa., started his own gunsmith establishment in 1751 in the same town, making arms principally for the Indian traders. In 1755 William Henry was armorer to the colonial forces with the Braddock Expedition, and saw more military service in 1758, with the Forbes Expedition against Pittsburg.

William Henry I is known to have furnished arms to the Continental troops in 1776, and was authorized to make muskets for the State of Pennsylvania in September, 1777. He died in 1786 at the age of 57, but, since the gun shop was not mentioned in his will, it is quite likely that his son, William II, (who had been trained under Andrew Albright of Lititz, Pa.) took over the management of the family plant, and moved the equipment when he established himself as an arms maker in Nazareth, Pa., in 1780. There he is known to have trained his sons, John Joseph (initials often written I. I. Henry) and William III, (younger of the two) in the gun manufacturing trade.

On June 30, 1808, William Henry II and his son John Joseph obtained a government contract of five years duration for 10,000 stands of arms, of which it is recorded that 4,246 were delivered by Oct. 7, 1812, and so presumably the entire contract was filled in due time.

The Nazareth plant being inadequate to fill the contract, in April, 1812, the younger son, William III was sent to Boulton, a mile south of Jacobsburg, to build a dam, a factory and workmens' houses on land owned by the Henrys and so found the Boulton Gun Works.

Though the Henry plant was at Boulton, the firm maintained a separate plant in Philadelphia, where John Joseph Henry had his offices and salesrooms and where the greater part of the firm's business was transacted. On Feb. 9, 1815, J. J. Henry obtained an additional government contract for 2,277 muskets.

About 1822 William Henry III sold out his interests to John Joseph Henry, who moved to Boulton and took into partnership his son James, some of the arms produced by the firm thereupon being marked "J. J. HENRY & SON".

During the War of 1812, John Joseph Henry was active in charge of production and repair of public arms for the Committee of Defence of Philadelphia. Upon the death of J. J. Henry in 1836, the works passed to his son James Henry.

J. J. HENRY (Boulton) FLINTLOCK PISTOL, MODEL 1826 TYPE

Illustrated—Fig. 4, Plate 13.

Caliber .54, using a half ounce spherical ball. Eight-and-one-half inch round, smoothbore barrel. Total length thirteen-and-one-half inches. Weight 2 pounds, 4 ounces. Iron mountings. The barrel is held by a spring fastened, single band, formed like the lower band of a musket, and carries a knife-blade brass front sight at the muzzle. The iron barrel tang, carrying a large open rear sight, extends down the rounded butt to meet a short extension of the iron butt cap. The full length, walnut stock forms almost a right angle to the butt; the front part of the stock extends up to the swivel of the steel ramrod. The lock-plate is flat and bevel-edged at the front, and the surface is slightly rounded at the rear. The brass pan is slightly tilted at the rear and is without a fence. Double-necked, rounded face hammer. The small end of the swivel ramrod is threaded inside to take a ball screw or a wiper head.

In the specimen illustrated the lock-plate is marked "J. J. HENRY BOULTON", in two lines between the hammer and the frizzen. The barrel is unmarked. The pistol was probably made for militia use. (See J. J. Henry, Boulton.)

C. KLINE Flintlock Pistol

Illustrated—Fig. 5, Plate 13.

Caliber .48. Nine-and-one-quarter inch round, smooth-bore barrel. Total length sixteen inches. Brass mountings. The barrel and thimbles are pin fastened. The walnut stock extends to the muzzle. The club-like stock has but little curve and ends in a shallow butt plate. Brass blade front sight. There is no rear sight or back-strap. The trigger guard forks at the rear about halfway up the loop. Flat goose-neck hammer. Iron pan with fence. Swelled hickory ramrod.

The lock-plate is marked "C. KLINE" between hammer and frizzen spring. The barrel is unmarked.

KUNTZ Flintlock Pistol

Illustrated—Fig. 1, Plate 14.

Caliber .48. Seven-and-five-eighths inch round, *brass* smoothbore barrel, octagonal to the rear. Total length twelve-and-three-quarters inches. Brass mountings. The barrel and the two thimbles are pin fastened. The full length walnut stock ends in a brass end-cap, flush with the muzzle. The barrel carries a brass blade front sight; there is no rear sight. The bird-head shaped butt has a shallow butt cap. There is no back-strap. The brass trigger guard is reinforced at the forward end and forks at the rear to complete the narrowed oval. Flat goose-neck hammer; iron pan with fence. The frizzen ends in a graceful curl. Hickory ramrod.

The barrel of the pistol illustrated is marked "KUNTZ PHILAD". The imported British lock-plate is stamped "T. KETLAND & CO."

The pistol was made by Jacob Kuntz, a Philadelphia gunsmith.

LINDSAY PERCUSSION PISTOL
Illustrated—Fig. 2, Plate 14.

Caliber .45, two shot. Eight-and-one-quarter inch semi-octagonal, smoothbore barrel. Total length twelve inches. Weight 2 pounds, 8 ounces. A brass blade front sight is mounted on the barrel. The rear sight is a notch, cut in the brass frame. Not equipped with a ramrod.

The pistol fired two charges, loaded from the muzzle, one in front of the other into a single barrel. The arm was discharged by a lock mechanism, consisting of two centrally hung percussion cap hammers operated by a single trigger, the right hammer falling first on the right cone and discharging the forward load through a channel leading to the forward charge. The rear load was likewise fired by the left hammer, by successive pressure on the trigger, the flash communicating from the cone directly to the charge.

In the specimen illustrated the barrel is marked in front

of the cones, "LINDSAY'S YOUNG AMERICA PAT-ENTED OCT. 9, 1860". Serial number 64 is marked on the bottom of the barrel and on the brass frame in front of the trigger guard.

The mechanism was the invention of John P. Lindsay, a former employee of the Springfield Armory. It is believed that about one hundred of these large caliber pistols were made. The arms were submitted to the government but were turned down. However one thousand Lindsay two-shot muzzle-loading muskets were purchased by the War Department in Aug. 1864. Legend has it that the two-shot musket was designed to surprise Indians, who had wiped out a command in which Lindsay's brother was a soldier. The Indians drew the fire of the troops equipped with the usual single shot muskets, and then charged in overwhelming numbers before the muzzle-loaders could be reloaded.

The principle of multi-firing arms operating by discharging successively, charges loaded in front of one another, was known for some hundreds of years before the Lindsay patent of 1860. A four-shot, superposed charge, single barrel flintlock pistol made by Golcher, is illustrated by Mr. Sawyer in Plate 28 of his volume "Firearms In American History".

However, though Mr. Lindsay's trigger mechanism was ingenious and reliable, the system of superposed charges was a failure; — the channel leading to the forward charge was blocked with fouling after a number of shots, and often the flash of the forward discharge leaked past the bullet of the rear charge, which was supposed to act as a gas check, resulting in a simultaneous discharge of both loads.

MARSTON Breech-loading Percussion Pistol

Illustrated—Fig. 3, Plate 14.

Caliber .36, single shot. Five-and-three-quarters inch barrel, round at the front, octagonal to the rear and rifled with six grooves. Total length ten-and-three-quarters inches. Weight 1 pound, 12 ounces. A brass, blade front sight and a V-notch iron, open rear sight are mounted on the barrel. Oval iron trigger guard. The side hammer and the silver plated, bronze frame are engraved. Blue barrel; case-hardened hammer, trigger and operating lever. Walnut grips shellacked to a high finish.

The sliding breech-block is operated by a lever located in front of the grip, in which the lever is partly sheathed. Forward movement of the lever opens the sliding breech-block, and permits the loading of the chamber with a special cartridge. The barrel may be removed by releasing a pin in front of the percussion cone. The pistol used a cartridge made of a cardboard shell with a round, leather base, with a small hole punched through at the center to admit the flash from the cone to the gun cotton load.

The barrel of the specimen illustrated is marked "W. W. MARSTON PATENTED NEW YORK" on top and "CAST STEEL" on the right side of the octagonal part. Number 76 is marked on the barrel and the frame.

The pistol was patented by William W. Marston of New York City, June 18, 1850, Patent No. 7,443, and was manufactured in a number of barrel lengths, such as six, seven and eight-and-one-half inch. Pistols of later production were made with iron frames.

McK BROTHERS (Baltimore) PERCUSSION PISTOL

Illustrated—Fig. 5, Plate 14.

Caliber .60. Ten-and-three-eighth inch smoothbore barrel, round, for about the first third from the muzzle, and fluted on top, rearward to the breech. Total length sixteen inches. Brass mountings. The barrel and the two split, brass thimbles are pin fastened. The full length stock extends to within three-eighths of an inch of the muzzle, and is narrowed in front of the forward thimble possibly to facilitate the removal of the hickory ramrod. Into the butt end is set a small rosette shaped butt cap. There is no back-strap. The oval trigger guard forks at the rear to complete the oval. The cone is set into a cylindrical side lug.

The flat lock-plate is merely marked "McK BROTHERS BALTIMORE". The barrel is unmarked.

"McK" probably stands for McKim, Baltimore gun-smiths.

MEACHAM & POND FLINTLOCK PISTOL

Caliber .54. Eight-and-one-half inch round, smooth-bore barrel without sights. Total length about thirteen-and-one-quarter inches. Brass mountings. The barrel and thimbles are pin fastened. The full length walnut stock extends to within one-eighth of an inch of the muzzle. The plain butt is rounded and is without butt cap or back-strap. Plain loop trigger guard. Goose-neck hammer. Iron pan with fence. Swell-tipped hickory ramrod.

In the specimen described, the flat, bevelled-edge lock-plate is marked "MEACHAM & POND WARRANT-

ED" in three lines between the hammer and the frizzen. The end of the lock-plate behind the hammer is engraved.

The pistol is believed to have been made in Albany, N. Y., in the early eighteen-hundreds.

MILES Flintlock Pistol, MODEL 1808 Type
Illustrated—Fig. 5, Plate 14.

Caliber .58. Nine-and-three-quarters inch round, smoothbore barrel without sights. Total length fifteen-and-one-half inches. Weight 2 pounds, 6 ounces. Brass mountings. The barrel, trigger guard and the two thimbles are pin fastened. The full length stock extends to within one-eighth of an inch of the muzzle. The fish-tail shaped, rounded butt ends in a brass butt cap with rounded side extensions. There is no back-strap. The trigger guard forks at the rear to complete the oval. Flat, bevelled, double-necked hammer. Horizontal iron pan with fence. Hickory swell tipped ramrod. The band at the front of the stock of the pistol illustrated is a later addition to reinforce a break.

The barrel of the specimen illustrated is marked "MILES PHILADA P". The flat, bevelled-edge lock-plate is marked "MILES PHILA" in an oval, between the frizzen spring and the hammer.

Of the family of Miles, musket makers, John Miles, Sr., was born in London, England, in 1752; where his son John Jr. was also born in 1777. The Miles family emigrated to the United States about 1790, and settled in Philadelphia, where they lived at 500 North Second Street from 1790 until 1798. In 1805 John Jr. is listed in the Philadelphia

City Directory, as residing at 43 Chestnut Street, while John Miles, Sr. is shown living at 30 South Third Street in 1805-6-7.

John Miles, Sr., had Pennsylvania State contracts of 1798 and 1801 for a total of 4,000 Model 1795 muskets. Upon his death in 1808, his son moved to Bordentown, New Jersey, where he obtained on July 20, 1808, a contract for 9,200 muskets, of which 2,407 are on record as having been delivered by Oct. 7, 1812. Of the arms delivered by Miles under the 1808 contract, many parts were obtained from sub-contractors in and around Philadelphia.

There is no record available of pistol contracts awarded to Miles, father or son. Judging from the resemblance of this pistol to the Model 1808 North Navy pistol which was made according to a government pattern, probabilities are that the specimen described was made by John Miles, Jr., for militia use, or for sale to individual officers or privateers.

MILES FLINTLOCK PISTOL
Illustrated—Fig. 1, Plate 15.

Caliber .64, taking an ounce, spherical ball. Nine-and-three-eighths inch round, smoothbore barrel. Total length fifteen-and-one-quarter inches. Brass mountings. The barrel and thimbles are pin fastened. The full length walnut stock extends to within one-quarter of an inch of the muzzle. There are no sights. The sloping butt ends in a plain, brass butt cap with short, rounded side extensions. Plain loop trigger guard. Goose-neck hammer. Iron pan with fence. Hickory ramrod.

The flat bevelled-edge lock-plate is marked "MILES" and "CP" behind the hammer, and has two vertical grooves. The barrel is unmarked.

The CP indicates that this militia pistol was the property of the Commonwealth of Pennsylvania, whose arms were required to be so marked by par. 1, Section 1, of the Act of March 28, 1797: —

> ". . . and to be stamped or marked near the breech with the letters C.P. the locks to be upon the best construction, double bridled, on a flat plate and marked with the letters aforesaid . . ."

B. MILLS PERCUSSION PISTOL

Caliber .75. Ten - and - three - quarters inch round, smoothbore barrel. Total length sixteen - and - three - quarters inches. German silver mountings. The barrel is key fastened to the half-stock which is tipped by a German silver end-cap. The barrel has a rib extension from the end of the stock to the muzzle. The sloping rounded butt is without a butt cap or back-strap. Plain, iron trigger guard. Bolster type cone seat. The hickory ramrod is brass tipped.

The lock-plate of the specimen described is marked "B. MILLS HARRODSBURG KY." The barrel is unmarked.

The pistol was made by Benjamin Mills, a gunsmith of Charlottesville, North Carolina, who served with Morgan's Rifles in the War of the Revolution, and settled in Harrodsburg, Kentucky, about 1790, where he made arms until about 1815. The settlement of Harrodsburg was founded by Col. James Harrod.

Benjamin Mills is reputed to have armed Colonel Richard M. Johnson's regiment of mounted Kentucky riflemen who decisively defeated the British and their Indian allies under General Proctor in the battle of Thames River, near Moravian Town, Canada, on October 5, 1813. Tecumseh was among the slain, and as a result of this battle his Indians deserted the British cause.

P. & D. MOLL Flintlock Pistol
Illustrated—Fig. 2, Plate 15.

Caliber .38. Eight-and-three-eighths inch octagonal *brass* barrel, unique in that the specimen illustrated was rifled with eight grooves. Total length fourteen inches. Weight 2 pounds, 2 ounces. Brass mountings. The barrel and two thimbles are pin fastened. On the barrel is mounted a brass, knife-blade front sight and an open, iron, V-notch rear sight. The full length maple stock extends to the muzzle. The curved-in, rounded butt ends in a brass butt cap which curves at the rear, up the stock. Goose-neck hammer. The iron pan is rounded at the bottom, has a fence and was forged integral with the lock-plate. The brass trigger guard forks at the rear to complete the oval. Hickory swell-tipped ramrod.

The barrel is marked "P. & D. MOLL HELLERS-TOWN". The flat, bevelled-edge lock-plate of the imported lock is marked "LONDON WARRANTED" between the hammer and the frizzen spring.

The maple stock of the pistol is "tiger striped" after the fashion of Kentucky rifles, an effect procured by burning a heavy tarred twine wrapped around the stock in the rough: the heat hardened wood under the burning twine producing on the finished stock the effect of curly maple in striped pattern.

The pistol was probably made about 1812, by Peter and David Moll of Hellerstown, Penna. The Molls specialized in brass barreled pistols and were expert in artificially graining maple stocks.

Legend has it that the Molls made a sufficient number of these rifled, brass barreled pistols to equip a troop of cavalry in the War of 1812.

I. PERKIN FLINTLOCK PISTOL
Illustrated—Fig. 3, Plate 15.

Caliber. 62. Eight-and-seven-eighths inch round, *brass,* smoothbore barrel. Total length fourteen-and-one-eighth inches. Brass mountings. The barrel and thimbles are pin fastened. The full length walnut stock extends to within one-quarter of an inch of the muzzle. The rounded fish-tail shaped butt ends in a shaped and engraved butt cap with long, graceful side extensions reinforcing the stock. There are no sights. The brass trigger guard forks at the rear to form a well curved oval. Flat goose-neck hammer engraved with a border line and ending in a curl tip. Iron pan with fence. The frizzen tip ends in a curl. Hickory ramrod.

The barrel is flattened at the breech and on the flat is marked "I. PERKIN". The lock-plate is also marked "I. PERKIN", between the hammer and the frizzen.

PERRY Breech-loading Percussion Pistol
Illustrated—Figs. 4 & 5, Plate 15.

Caliber .52, single shot. Six-and-three-sixteenths inch round barrel, rifled with six grooves. Total length twelve-and-three-quarters inches. Weight 2 pounds, 15 ounces. Brass blade front sight; V-notch rear sight, is cut into the barrel over the breech. There is no forearm. Oil finished walnut grip.

Side hammer. Lowering the trigger guard tilts the breech-block upwards, and permits loading with paper cartridge, or loose powder and ball. The breech-block has a projecting bevelled ring which fits into the chamber and forms a gas tight joint. A brass tube capping magazine extends upward through the butt, and feeds a percussion cap automatically to the cone when the breech-block is closed. The pistol was made in a number of barrel lengths and both with and without automatic capping device.

The specimen illustrated is marked on the top of the breech-block "A. D. PERRY PATENTED" and "PERRY PATENT FIRE ARMS CO. NEWARK N. J."

The mechanism of the pistol was patented by Alonzo D. Perry, Patent No. 12,244, January 17, 1855.

Perry pistols were offered to the government in both army and navy sizes, but were rejected. About 1855 the Perry Patent Fire Arms Company became involved in financial difficulties and failed.

POND Flintlock Pistol

Caliber .56. Nine inch octagonal, *brass,* smoothbore barrel. Total length fifteen inches. Weight 3 pounds.

Brass mountings. The barrel is key fastened to the full length applewood stock, which extends to the muzzle. Pin fastened thimbles. The rounded butt is reinforced by a plain brass butt cap. There is no back-strap. The brass trigger guard forks at the rear to complete the oval. Brass cone front sight; there is no rear sight. Goose-neck hammer; iron pan with fence, the frizzen spring has a roller. Hickory ramrod.

The flat, engraved lock-plate is marked "POND & CO." between the hammer and the frizzen spring. The barrel is marked "ALBANY".

RICHMOND-VIRGINIA FLINTLOCK PISTOL
Illustrated—Fig. 1, Plate 16.

Caliber .54, taking a half ounce spherical ball. Ten inch round, smoothbore barrel without sights. Total length sixteen-and-one eighth inches. Weight 2 pounds, 13 ounces. Brass mountings. The barrel is key fastened to the walnut stock and has a rib extension underneath, from the end of the half stock to the swivel of the steel, swivel ramrod. The rounded butt ends in a plain, brass butt cap which has an extension on each side, set flush with the wood and extending upward, to reinforce and decorate the grip. There is no back-strap. The hammer is double-necked, flat with bevelled-edges. Iron pan with fence. The brass trigger guard forks at the rear into a curl to complete the oval. The ramrod thimble and the barrel reinforcing band are brass.

In the specimen illustrated the flat, bevelled-edge lock-

plate is marked "VIRGINIA" between the hammer and
the frizzen, and "RICHMOND" in a curve, and "1813"
behind the hammer. Specimens similar to this pistol are
also known marked "RICHMOND" in a curve in front
of hammer, and dated (1815) behind the hammer.

The pistol was made at the Virginia Manufactory, an
armory established by the State of Virginia to provide
arms for the state militia. (See Virginia Manufactory).
The arms of this model were manufactured from 1812 to
1816, and were made in close resemblance of the Model
1806 Harpers Ferry pistols, except for the improvement
of the swivel ramrod.

ROGERS & BROTHERS Flintlock Pistol

Caliber .54. Eight-and-one-half inch semi-octagonal,
smoothbore barrel. Total length about fourteen inches.
Brass mountings. The barrel is key fastened with two
keys. Pinned thimbles. The striped, maple stock of this
specimen extends to within one-eighth of an inch of the
muzzle and ends in a brass end-cap. The bird-head shaped,
checked butt is reinforced by a shallow butt cap. The trigger guard forks at the rear. A large, brass, blade front
sight is mounted on the barrel. There is no rear sight.
Goose-neck hammer; iron pan with fence. The frizzen
spring is equipped with a roller. Hickory, brass tipped
ramrod.

The flat, engraved lock-plate of the specimen described
is marked "ROGERS & BROTHERS" and "WAR-
RANTED".

It is possible that the Rogers & Brothers pistols were made at Valley Forge between 1814 and 1825.

John, Charles and Evan Rogers, hardware merchants, are listed in the Philadelphia City Directories from 1805 to 1846, principally at 52 High Street. John Rogers also owned the Valley Forge which he purchased in 1814. On March 21, 1821, John Rogers and Brooke Evans took over a defaulted government contract for 10,000 muskets which had been awarded to Alexander McRae on July 28, 1817. On January 1, 1825, John Rogers obtained a con-tract for 5,000 muskets, which he probably shared with William L. Evans, a practical gunmaker who managed the Valley Forge works.

The Valley Forge descended to a nephew, Charles H. Rogers, who lived at the Forge. In 1839 the Valley Forge Gun Factory was partially destroyed by a freshet, and was completely destroyed in 1843. The estate then went to female descendants until bought by the State of Pennsyl-vania for a State Park.

JOHN RUPP Flintlock Pistol

Caliber .47. Eight-and-one-half inch round, smoothbore barrel, octagonal for the rear half. Total length thirteen-and-three-quarters inches. Silver mountings. The barrel is key-fastened with two keys. The thimbles are pinned. The full length, red-finished maple stock ends one-eighth of an inch from the muzzle. The butt is bird-head shaped, checked and ends in a shallow silver butt cap. The silver trigger guard forks at the rear. Brass blade front sight;

there is no rear sight. Goose-neck hammer; iron pan with fence. Swell tipped hickory ramrod.

The flat, bevelled-edge lock-plate is unmarked except for two vertical lines behind the hammer. The barrel is marked "JOHN RUPP".

John Rupp, a Pennsylania gunsmith was established at Ruppville, near Allentown about 1780.

SHARPS BREECH-LOADING PERCUSSION PISTOL
Illustrated—Fig. 2, Plate 16.

Caliber .38, single shot. Six-and-one-half inch round, tapered barrel rifled with 5 grooves. Total length eleven inches. Weight 2 pounds. A brass blade front sight is mounted on the barrel; an open V-notch rear sight is in the frame. The frame and back-strap are steel. Two piece walnut grips. The pistol described was made without a fore-end.

The pistol is equipped with a side hammer and a Lawrence pellet primer. The cylindrical breech-block is operated by a lever forming the trigger guard. Lowering the guard opens the chamber for loading with a combustible cartridge.

The pellet priming device embodying the Sharps patent of 1852, and the Lawrence patent of 1856, consisted of a tube seated in the frame of the arm, from which tube priming pellets were automatically fed to the cone by the action of cocking the hammer.

The specimen illustrated is marked on the left side of the frame "C. SHARPS & CO. RIFLE WORKS PHILA. PA. C. SHARPS PATENT 1848".

The mechanism of the arm was patented by Christian Sharps, Sept. 12, 1848, Patent No. 5,763. The pistol was manufactured by C. Sharps & Co. of Philadelphia (1859-1863). These arms were made in various calibers, in different barrel lengths and with minor variations in construction, such as with or without fore-ends, and with wood or metallic grips. Some were marked "SHARPS PATENT ARMS MFᴇᴅ. FAIR MOUNT PHILA. PA." In others the patent dates were given "C. SHARPS PATENT 1848-52".

A few were made in caliber .44, and were submitted to the U. S. military and naval services, but were rejected. A later modification to cartridge type was also declined.

In 1859 "C. Sharps & Co." consisted of Christian Sharps, N. H. Bolles and J. B. Eddy. The firm was located in West Philadelphia (Fairmount). In 1863 the company was combined with Wm. C. Hankins, rifle manufacturer, to form the "Sharps & Hankins" rifle and pistol manufacturing establishment.

SHULER Fʟɪɴᴛʟᴏᴄᴋ Pɪsᴛᴏʟ, MODEL 1808 Tʏᴘᴇ

Caliber .54, taking a half ounce spherical ball. Nine-and-five-eighths inch round, smoothbore barrel. Total length sixteen inches. Weight 2 pounds, 5½ ounces. Brass mountings. The barrel and thimbles are pin fastened. The full length walnut stock extends to within one-sixteenth of an inch of the muzzle, and is reinforced by a brass band. The sloping, fish-tail shaped butt ends in a brass butt cap with short, rounded side extensions. There is no back-strap. Brass blade front sight; there is no rear sight. Plain brass loop trigger guard. Double-necked hammer; bevelled iron pan with fence. Hickory ramrod.

The barrel of the specimen described is marked "SHULER". The flat, bevelled-edge lock-plate is marked "US" behind the hammer.

The pistol is believed to have been made by J. Shuler of Liverpool, Penna., for militia use and dates to about 1812. It is similar in appearance to the Henry pistol of the Model 1808 type.

SWEITZER Flintlock Pistol, MODEL 1808 Type

Caliber .54, taking a half ounce spherical ball. Ten- and-seven-sixteenths inch round, smoothbore barrel. Total length sixteen inches. Weight 2 pounds, 6 ounces. Brass mountings. The barrel and thimbles are pin fastened to the full length walnut stock, which extends to within one-sixteenth of an inch of the muzzle. The butt is reinforced by a brass butt cap with short, rounded side extensions. There is no back-strap. A brass, blade front sight is mounted on the barrel; there is no rear sight. The brass trigger guard forks at the rear to complete the loop. Straight trigger. Flat, bevelled-edge, double-necked hammer with a curl at the top. The iron pan has a fence, and is forged separately from the lock-plate. Hickory ramrod.

The flat, bevelled-edge lock-plate is marked "SWEITZER & CO." in two lines between the hammer and the frizzen spring. The barrel is marked with an eagle head and letters "CT", all in the same oval.

It is believed that the firm "Sweitzer & Co.", makers of the pistol, is identical with "Daniel Sweitzer & Co.", Pennsylvania gunmakers of Lancaster, Pa., who in 1808 announced the opening of their . . . "gunlock factory, west of the court house, on the road to Millerstown".

VIRGINIA MANUFACTORY Flintlock Pistol
Illustrated—Fig. 3, Plate 16.

Caliber .69, taking an ounce spherical ball. Twelve-and-one-eighth inch round, smoothbore barrel. Total length seventeen inches. Weight 3 pounds, 6 ounces. Iron mountings. The barrel is held by an iron double band with a brass, blade front sight mounted on the rear strap. There is no rear sight. The rounded butt is reinforced by an iron back-strap running from the tang to the butt cap. The iron butt cap has short rounded extensions reaching into the stock on both sides. The full length walnut stock extends to within about half an inch of the muzzle. Steel ramrod. Goose-neck hammer. The pan is iron, with fence, and is forged integral with the lock-plate.

In the specimen described the barrel is marked "P (W) REG. 3 V 1810". The flat, bevelled-edge lock-plate is marked between the hammer and the frizzen spring "VIRGINIA Manufactory", the latter word in script. Behind the hammer is stamped "RICHMOND 1806" in a circle.

In other specimens the stamping "RICHMOND" and the date are in two curved lines. Such a specimen, dated 1807, is illustrated.

These arms are known dated from 1805 to 1809. Being hand made there is some variation in barrel lengths which run from eleven-and-three-quarters to twelve-and-one-quarter inches, the total length of the arm and its weight, varying in proportion.

The establishment of the Virginia Manufactory was authorized by an Act of 1797, Virginia Legislature, to found an armory for the manufacture of arms to equip the state militia.

Production of arms began five years later, in 1802, and continued until 1820, when manufacture was discontinued and the plant was converted into a school. In 1860 the armory was rehabilitated, and resumed the manufacture of arms for the Confederacy until the close of the Civil War.

In connection with pistol production at the Virginia Manufactory the following report of Dec. 18, 1806, from John Clarke, the superintendent, to the Governor of Virginia, is of interest: —

"In compliance with your desire as stated in your letter of the 13th instant, I have to state that the arms for all the Troops of Cavalry ordered to be armed by the Executive, have been sent to them except Capt. Riddick's Troop and the Troop commanded by Capt. Muse; the arms for which are not yet sent from the armory, although they are stamped in the usual manner and are ready for delivery at any moment, which completes all the orders I have received for arming the cavalry. Besides which we have now on hand 173 Virginia manufactured pistols, . . ."

It was John Clarke who, on Jan. 19, 1804, suggested to the governor that the manufacture of pistols be undertaken to utilize: —

". . . many skelps and barrels refused on account of flaws and other defects, the best parts of which would make good pistol barrels . . . particularly if there is to be no difference between the calibre of the muskets and that of the pistols."

J. WALSH FLINTLOCK PISTOL
Illustrated—Fig. 4, Plate 16.

Caliber .54, taking a half ounce spherical ball. Eight inch *brass,* smoothbore barrel of tapering cannon shape, with a brass cone front sight on the muzzle band. Total length fourteen inches. Weight 1 pound, 15 ounces. Brass mountings. The barrel and the two ramrod thimbles are pin fastened. The full length walnut stock extends to within about half inch of the muzzle. The rounded butt is fish-tail shaped at the engraved butt cap, which has three inch extensions set flush with the stock and reaching upward and forward to reinforce the butt. The brass trigger guard has an ornamental forward extension, and forks at the rear to complete the oval. The rear branch continues under the stock to end just short of the butt cap. Curled tip, early type trigger. Ornamental brass side plate. The hammer is of the early goose-neck type. The brass pan is rounded at the bottom, has a fence and is cast integral with the lock-plate. The frizzen heel ends in a graceful curl. Swell tipped hickory ramrod.

The barrel is marked "J. WALSH PHILAD", in a panel outlined by an engraved design. The flat, bevelled, brass lock-plate is pointed at the rear, and is marked between the hammer and frizzen spring "J. WALSH", within an engraved design. The tasteful engraving is characteristic of all the mountings of the pistol.

James Walsh was one of a number of Philadelphia gunsmiths who in November, 1776, protested to the Pennsylvania Committee of Safety in regard to the high price of materials entering into gun making. In 1779 Walsh advertised, offering his gunsmith tools for sale.

A. H. WATERS & CO. PERCUSSION PISTOL
Illustrated—Fig. 5, Plate 16.

Caliber .54, taking a half ounce spherical ball and a charge of 50 grains of rifle powder. Eight-and-one-half inch round, smoothbore barrel, finished bright. Weight 2 pounds, 6 ounces. All mountings are iron. The cone is set on a side lug with a square end. The barrel carries a brass knife-blade front sight and is held to the stock by a single branch-band fastened on the left side through the forward side screw. A large, open rear sight is on the tang. A long butt cap extension forms the back-strap. The three-quarter length, oil finished, black walnut stock ends short of the swivel of the ramrod. The small end of the ramrod is threaded to take a ball screw or a wiper head.

In the specimen illustrated the barrel is marked "NWP". The flat, unbevelled lock-plate is marked "A. H. WATERS & CO. MILBURY MASS." and is dated "1844".

This A. H. Waters & Co., percussion pistol is believed to have been made for private sale to officers, or for militia use. Except for the fact that it was made on the percussion system, (not an alteration), in the manner of the Model 1842 percussion pistols, it is similar to the A. Waters and A. H. Waters & Co., flintlock pistols in all external details, except that the edges of the lock-plate are not bevelled. Specimens are also known equipped with brass trigger guards.

BUTT CAP
DERINGER
1808

UNITED STATES MARTIAL REVOLVERS

AND

AUTOMATIC PISTOLS

Chapter I.

MARTIAL PERCUSSION REVOLVERS

PERCUSSION REVOLVERS — An outline of Colt history — Combustible cartridges — Instructions for loading percussion revolvers —

ADAMS —
ALLEN & WHEELOCK —
ALSOP —
BEALS —
BUTTERFIELD —
COLT Modle 1839 —
COLT Model 1847 —
COLT Model 1848 —
COLT Model 1851 —
COLT Model 1855 —
COLT Model 1860 —
COLT Model 1861 —
COLT Model 1862 —
COOPER —
FREEMAN —
JOSLYN —
LEAVITT —
MANHATTAN —
METROPOLITAN —
PETTINGILL —
PLANT —
REMINGTON Model 1861 —
REMINGTON New Model —
REMINGTON-RIDER —
ROGERS & SPENCER —
SAVAGE-NORTH —
SAVAGE —
STARR —
UNION —
WALCH —
WARNER —
WESSON & LEAVITT —
WHITNEY —

UNITED STATES MARTIAL REVOLVER
AND AUTOMATIC PISTOLS

PART II.

Chapter I.

UNITED STATES MARTIAL PERCUSSION REVOLVERS

PERCUSSION REVOLVERS.

The principle of multi-firing arms was known many years before February 25th, 1836, when Samuel Colt of Hartford, Connecticut, obtained a patent for firearms firing successive charges presented before a single barrel, by means of revolutions of a cylinder containing the loads. Even that idea was not new: small-arms embodying the principle of a revolving breech had been made in the matchlock, wheel lock and snaphance ignition system constructions, and are to be found in the important museums of the world. A successful and ingenious, manually operated revolving cylinder, flintlock weapon was patented in Great Britain in 1818 by Elisha Collier of Boston, Mass., who after six years of experimentation went to England to manufacture and market his invention because of the high cost of labor, and consequently limited market, for his expensive weapons at home. But the Collier revolving pistol, like its predecessors, was unsuitable for military or even general use. It was complicated, delicate, expensive in construction, but its chief objection lay in the very nature of its cumbersome, outside, flintlock mechanism, which was apt to result in the unexpected discharge of several or all of the chambers at once.

The invention of the percussion ignition system permitted design that did away with the danger of simultaneous discharge of several chambers, and to Samuel Colt belongs the credit for perfecting the first practicable and working firearm equipped with a mechanically operated cylinder, and its subsequent improvement and development. In addition to the mechanical revolution of the breech, his patents covered a number of other vital points, such as placing the percussion cap cones on the back of the loading chambers, the separation of the cones by metal walls to prevent the ignition of chambers other than the one to fire, and a mechanical method of locking the cylinder at the moment of firing to insure the alignment of the chamber with the barrel. In fact his patents covered the important and basic principles of revolver construction so thoroughly, that it was not until the expiration of the Colt patents in 1856, that practical revolvers could be manufactured by others without infringing the Colt patents.

Upon the receipt of his United States and British patents, young Colt,—he was but about twenty-two at that time—succeeded in interesting a sufficient number of capitalists in his invention to organize the "Patent Arms Manufacturing Compay" at Paterson, New Jersey;—a stock company with a capitalization of about $230,000. Colt sold his patents rights to the new concern for a cash payment of some $6,000, royalties, and a percentage of the profits; Colt himself becoming an employee of the firm in the capacity of a salaried manager.

The first concealed trigger caliber .34 pocket revolvers
made at the Paterson works were an improvement over
the same arms made for Colt by Anson Chase of Hartford and Frederick Hanson of Baltimore. The first output
was shortly followed by a number of other models of various barrel lengths and calibers ranging from .22 to .50.
Like all new inventions the new arms were slow to "take"
in the conservative East, and a large part of the company's
stock of hand arms was purchased by enterprising Texan
arms dealers. Though the War of Texas Independence
was over, the frontier conditions were anything but peace-
ful, and the revolvers won instant popularity in the turbulent Lone Star Republic.

These first Paterson, or Texas Colts, as they are now
called among the collectors, were further improved after
1839 by newly patented features such as a hinged lever
rammer, simplified operating mechanism and facilities for
the rapid loading of the cylinder by special powder and
ball flasks and cap magazines. In the same year a heavier
martial revolver equipped with the conventional trigger,
trigger guard and a hinged lever rammer of the later percussion revolvers, was made for the use of the Texas
Rangers.

The lever ramrod and other improvements of 1839 are
ascribed partly to suggestions made to Colt by representatives of the Texas Government, who came to New York
to buy arms for the new republic. It is commonly reputed,
but not definitely established, that Captain Samuel H.
Walker of the Rangers was one of the agents, if not
indeed the sole representative. Be that as it may, these
revolvers, of which three hundred are said to have been

made, are referred to as Walker Colts in honor of the
Ranger captain, though his participation in their design
may be merely a legend.

In the meantime the financial affairs of the company
had not gone any too well. Despite their popularity in
Texas and their obvious superiority to single shot pistols,
the arms did not find a ready market in the East, the
center of population and purchasing power, and what was
more important, the Colt arms had been reported on
adversely by an army Ordnance Board convened at West
Point in March, 1837, to test new arms of the percussion
system. The ultra-conservative Board raised a number of
objections, among them being the temptation on the
soldiers' part to waste ammunition, the high cost of the
arms and excessive weight, and the valid reason of the
difficulty of maintenance of the relatively delicate arms
under field service conditions. But the real reason probably was that the arms were too new and untried, still in
the more or less experimental stage and in need of further
development through use. It is quite likely that another
objection was raised at the trials; — the old battle cry
heard in connection with every mechanical improvement
which increases the rapidity of fire;—"how are you going
to supply 'em with all the ammunition?"

In order to demonstrate the superiority of his repeating
arms, Colt even took a consignment to Florida, where the
United States was engaged in a small war with the Seminole Indians. The arms were readily sold to individual
army officers, but the long journey merely resulted in a
trifling order for fifty carbines through the interest of Lt.
Col. Harney. Later, in 1841, this purchase was followed

by an additional order for one hundred and sixty Model 1839 carbines for field trial at the Dragoon School at Carlisle Barracks, Penna.; but this small order could not compensate for a general lack of sales, and lacking public and government support the firm failed in 1842; the plant and the machinery were sold, and the patent rights eventually reverted to the inventor.

The number of arms manufactured at Paterson during the operations of the Patent Firearms Manufacturing Company is unknown. The theory has been advanced by Mr. Chas. T. Haven that by comparison of the number of known existing specimens of the Paterson Colt revolvers with the relatively few remaining Whitneyville Colts, Model 1847, of which a thousand are known to have been made, the total number made at Paterson must have been about five or six thousand. While this number does not seem excessive, keeping the output down to a modest figure of less than one hundred per month, there is one objection to Mr. Haven's method of estimation, and that is the fact that Model 1847 revolvers were issued to troops and saw hard service, and since the army custom has been to destroy unserviceable and condemned small-arms to prevent further use, but few of these arms are left of the thousand made and issued to the service.

After the failure of his company Samuel Colt occupied himself with other affairs until 1846, when the Mexican War created a demand for revolvers, and Samuel Colt was approached by the government to resume the manufacture of Colt arms. Colt's old friend, Captain Walker, now of the United States Army, was the emissary for the government in this transaction. Assured of a large order,

Colt contracted with Eli Whitney for the manufacture of the revolvers at the latter's armory at Whitneyville, Connecticut. Tradition has it that Colt was unable to locate a specimen of his martial 1839 revolver as made for the Texas Rangers, and designed a new arm patterned from memory after the "Walker".

The first order for one thousand revolvers was received in January, 1847, and was produced at Whitneyville. A second order for a thousand followed on November 2, 1847, and on the strength of that order Colt equipped a factory at Hartford, Conn., and completed the second thousand, known as Model 1848, without outside assistance. This model was also known as the "Dragoon" or "Holster Pistol", the word revolver not having come into general usage.

Further orders followed, and Colt engaged Elisha K. Root as factory superintendent, and devoted himself to the management of the prospering business. Though the Mexican War had ended, the westward trek and discovery of gold in California gave a new impetus to arms manufacturing, resulting in the enlargement of the Colt production facilities in a new plant at Hartford.

While the Model 1848 revolvers continued to be made with some minor variations, a new caliber .36 navy model was brought out in 1851, in addition to smaller caliber pocket revolvers introduced in 1848. Solid frame, side-hammer revolvers designed by Root were brought out in 1855, mostly in smaller calibers. The next martial revolver was the streamlined Army Model of 1860, followed by a companion Navy Model of 1861, the last of the Colt martial percussion revolvers.

Colt percussion revolvers similar to the American made models, except for steel trigger guards and butt-straps, were also produced in England at the London branch from 1853 until 1857, when the English armory was discontinued.

According to the Colt Fire Arms Company, during the Civil War the Colt Armory furnished the government with 386,417 revolvers, about 7,000 revolving rifles and carbines and 113,980 muzzle-loading rifle muskets. Col. Colt,—he had been commissioned Colonel in the Connecticut militia,—died January 10th, 1862, his end probably hastened by his labors in connection with supplying the government with arms in war emergency. Upon his demise Mr. Root became the president of the company.

The basic Colt patents issued in 1836 did not expire until 1856 due to an extension granted because of the cessation of manufacture between 1842, when the original company failed, and 1847, when Colt resumed the manufacture of arms under his patents.

With the expiration of Colt patents, the westward migration and the Civil War created a tremendous demand for arms, and a large number of firms brought out a host of revolvers, most of them embodying some Colt principles, while some, like the Manhattan and Metropolitan, were downright imitations in construction and design.

About 374,000 revolvers of many different makes were purchased by the United States during the Civil War, at an approximate cost of six million dollars. In the early days of the war the domestic purchases were supplemented by foreign arms, such as Adams, Lefaucheaux, Le Mat, Perrin, Raphael and others, but these were discarded for American arms as soon as the industry was able to supply the needs of the army and navy. Of these American made arms, the martial percussion revolvers, Colt and other makes, are described in this chapter.

The advent of the revolver increased the value of cav-alry. In addition to possible shock action, the value of cavalry in combat was to be found in its mobility and ability to displace its fire power quickly from one locality to another. Therefore its most effective employment was under conditions which not only permitted the fullest use of its mobility, but also the use of maximum fire power.

The issue of revolvers to the cavalry increased its fire power as compared to the infantry, which was still armed with muzzle-loaders. The advantage of increased fire power added to its mobility, enhanced the value of the mounted service until recent years, when the issue of repeating arms to the infantry and subsequent motoriza-tion caused a decline in the importance of cavalry as a combat branch.

In the following chapters are also included revolvers and automatic pistols, which though not adopted or pur-chased by the government, are definitely of martial type, and were produced in the hope of acceptance by the mili-tary and naval services.

THE COMBUSTIBLE CARTRIDGE.

Though the percussion revolvers could be loaded with loose powder and conical or round ball, the standard and convenient load was a self-consuming, combustible cart-ridge consisting of a charge of powder contained in a small paper or thin skin pouch, which was glued around the circumference of a conical bullet. The material of the pouch or bag, whether paper or skin, was usually prepared with nitre, and was consumed in the explosion with practically no residue. These combustible cartridges were sometimes made with a preserving glazed linen cover for the protection of the frail powder bag, the cover being removed before loading. Some makes were packed in small, hollow wooden containers, usually six cartridges to a case.

DIRECTIONS FOR LOADING A PERCUSSION REVOLVER.

The following instructions are copied from a label inside a Colt revolver case:

Directions for Loading Colt's Pistols.

First explode a cap on each nipple to clear them from oil or dust, then draw back the hammer to half-cock, which allows the cylinder to be rotated; a charge of powder is then placed in one of the chambers, keeping the barrel up, and a ball with the pointed end upward, without wadding or patch, is put into the mouth of the chamber, turned under the rammer, and forced down with the lever below the surface of the cylinder so that it can not hinder its rotation (care should be used in ramming down the ball so as not to shake out the powder from the chamber, thereby reducing the charge.) This is repeated until all the chambers are loaded. Percussion caps are then placed on the nipples on the right of the lock-frame, when, by drawing back the hammer to full cock, the arm is in condition for a discharge by pulling the trigger; a repetition of the same motion will produce like results, viz.: six shots without reloading.

N.B. ** Fine grain powder is best. Soft lead must be used for the balls. The cylinder is not to be taken off when loaded. To carry the arm safely when loaded, let down the hammer on one of the pins between each nipple, on the end of the cylinder.

Directions for Loading With Colt's Foil Cartridge.

Strip the white case of the cartridge, by holding the bullet end and tearing it down with the black tape. Place the cartridge in the mouth of the chamber of the cylinder, with the pointed end of the bullet uppermost, one at a time, and turn them under the rammer, forcing them down with the lever below the surface of the cylinder so they can not hinder its rotation.

To insure certainty of ignition, it is advisable to puncture the end of the cartridge, so that a small portion of gunpowder may escape into the chamber while loading the pistol.

ADAMS Navy Percussion Revolvers.   (MASS ARMS CO.)
Illustrated—Fig. 1, Plate 17.

Caliber .36, five shot, double action. Six inch octagonal barrel rifled with three grooves. Total length eleven-and-one-half inches. Weight 2 pounds, 9 ounces. The cylinder is one-and-fifteen-sixteenths inches long. A steel blade front sight is set into the barrel; the V-notch rear sight is on the frame: Oval, iron trigger guard. The checkered walnut grip has a hole bored through to take a lanyard.

The rammer is operated by a loading lever attached to the left side of the barrel. On the right side of the frame is a safety lock. The revolver used a self-consuming, combustible cartridge, though loose powder and ball could be used. All metal parts were blued.

The revolver is marked on top of the frame "MANUFACTURED BY MASS. ARMS CO. CHICOPEE FALLS". The frame is also marked on the right side "PATENT JUNE 3, 1856", and on the left side

"ADAM'S PATENT MAY 3, 1858". The loading lever is marked "KERR'S PATENT APRIL 14, 1857".

The revolver mechanism was patented in England in 1853. During the Civil War these revolvers were manufactured at Chicopee Falls by the Massachusetts Arms Company. English made Adams revolvers were also purchased by both the United States and the Confederate governments during the Civil War.

The Massachusetts Arms Company was organized by the heirs and kin of Edwin Wesson for the manufacture of revolvers under the Wesson patents. (See Wesson & Leavitt revolver).

ADAMS ARMY PERCUSSION REVOLVER.

Similar in design and construction to the Navy Model except that the caliber was .44 (This arm reputed made. No specimen was seen in any of the important collections examined.)

ALLEN & WHEELOCK ARMY PERCUSSION REVOLVER.
Illustrated—Fig. 2, Plate 17.

Caliber .44, six shot, single action. Seven-and-one-half inch round barrel, partly octagonal towards the breech and rifled with six grooves. Total length thirteen-and-one-quarter inches. Weight 2 pounds, 14 ounces. Center hammer with a V-shaped notch rear sight cut in the lip. A low blade, brass front sight is dovetailed in at the muzzle. The cylinder is one-and-fifteen-sixteenth inches long.

The trigger guard when released by a catch on its underside, drops, and by action of geared teeth operates the rammer to seat the bullet in the cylinder. The cylinder may be removed by pressing a spring in front of the frame and sliding the cylinder pin forward. The barrel, frame and cylinder are blued, the trigger guard and hammer are case-hardened in mottled colors. Varnished walnut grips. The revolver fired a self-consuming combustible cartridge, or could be loaded with loose powder and ball.

The specimen illustrated is marked on the left of the barrel "ALLEN & WHEELOCK WORCESTER MASS. U.S. ALLEN'S PT'S JAN. 13, DEC. 15, 1857. SEPT. 7, 1858". The cylinder is stamped "98".

The revolver was patented by Ethan Allen on Jan. 13, 1857, Pat. No. 16,367, Dec. 15, 1857, Pat. No. 18,836 and Sept. 7, 1858, Pat. No. 21,400. It was manufactured by Ethan Allen and his brother-in-law Thomas P. Wheelock at Worcester, Mass., from 1856 to 1865.

One hundred ninety-eight of these army revolvers were purchased by the War Department from Wm. Read & Son on Dec. 31, 1861. A total of about 500 Allen & Wheelock revolvers in both army and navy sizes are believed to have been purchased by the government during the Civil War.

After the introduction of the metallic cartridge the Allen & Wheelock revolver was changed into a cartridge arm by substituting a different cylinder, changing the hammer and adding a loading gate.

The firm Allen & Wheelock consisted of Ethan Allen, one of the early American arms makers, and his two brothers-in-law, Charles T. Thurber and Thomas P. Wheelock. The firm was known as Allen-Thurber & Co., until 1857, when Thurber retired and the name was changed to Allen & Wheelock. Their manufactory was located at Grafton, Mass., from 1838 until 1842, when they moved to Norwich, Conn. In 1847 the firm moved its establishment to Worcester, Mass.

Mr. Wheelock died in 1863, and in 1865 Allen's two sons-in-law, S. Forehand and H. C. Wadsworth were admitted to the firm. In 1866 the firm became Ethan Allen & Co. Mr. Allen died in January, 1871, and the firm continued under the name Forehand & Wadsworth.

ALLEN & WHEELOCK NAVY PERCUSSION REVOLVER.
SIDE-HAMMER MODEL.
Illustrated—Fig. 3, Plate 17.

Caliber .36, six shot, single action. Eight inch octagonal barrel rifled with six grooves. Total length thirteen-and-one-half inches. Weight 2 pounds, 6 ounces. The cylinder is one-and-twenty-seven-thirty-seconds inches long. A German silver blade is mounted on the barrel; the V-notch rear sight is in the frame. Blued finish.

The side hammer is on the right side. The trigger guard when released forms the loading lever and operates the rammer by geared teeth. The cylinder may be removed by withdrawing the cylinder pin from the rear. Like the army model, the navy revolvers were noted for the solid frame and a dependable rotating system which brought the chambers in exact alignment. The barrel, frame and cylinder were blued, the trigger guard and the hammer were case-hardened. The revolver used a combustible cartridge, or could be loaded with loose powder and ball.

The specimen illustrated is marked "ALLEN & WHEELOCK WORCESTER MASS. U.S. ALLEN'S PTS. JAN. 13, DEC. 15, 1857, SEPT. 7, 1858". The cylinder is engraved with a forest scene and animals. The revolver number 62, is stamped inside the loading lever.

This first model navy revolver was made in various barrel lengths, such as six, and seven-and-one-quarter, and later was made in a center hammer model similar in appearance to the army revolver.

ALSOP NAVY PERCUSSION REVOLVER.
Illustrated—Fig. 4, Plate 17.

Caliber .36, five shot, single action. Four-and-one-half inch octagonal barrel rifled with five grooves. Total length nine-and-seven-eighths inches. Weight 1 pound, 4 ounces. Round cylinder. Sheath trigger. A brass, cone front sight is mounted on the barrel, the V-notch rear sight is in the frame. Blued barrel frame and cylinder; case-hardened loading lever and hammer. Walnut grips.

When the hammer is cocked, the cylinder revolves and actuated by a cam, is moved forward against the barrel, forming a gas tight joint. Conventional loading lever. The revolver used a self-consuming, combustible cartridge, or could be loaded with loose powder and ball. Except for the sheath trigger, and the cylinder, revolved by the action of the hammer, the arm resembled the North Navy revolver.

The specimen illustrated is marked on the barrel "C. R. ALSOP MIDDLETOWN CONN. PATENTED

JULY 17th, AUGUST 7th, 1860, MAY 14th, 1861". The cylinder is marked "C. R. ALSOP PATENTED NOV. 26th, 1861".

The revolver was patented by Charles Alsop of Middletown, Connecticut, on the dates as above, Patent Numbers 29,213, 29,538 and 32,333.

This arm was also made in a longer, ten-and-three-quarter inch model with a five-and-three-eighths inch barrel equipped with fluted or round cylinders, and is believed to have also been made in caliber .44, army model.

BEALS NAVY PERCUSSION REVOLVER.
Illustrated—Fig. 5, Plate 17.

Caliber .36, six shot, single action. Seven-and-one-half inch octagonal barrel rifled with five grooves. Total length thirteen-and-three-eighths inches. Weight 2 pounds, 10 ounces. The cylinder is two inches long. A low blade, brass front sight is dovetailed into the barrel at the muzzle. The rear sight is in the frame. Oval brass trigger guard.

The cylinder is removed by lowering the conventional loading lever, and withdrawing the cylinder pin to the front. Blued finish. This Beals revolver has no intermediate hammer rest recesses cut into the cylinder. The revolver fired a self-consuming combustible cartridge, or could be loaded with loose powder and ball.

The specimen illustrated is marked on the barrel "BEALS PATENT SEPT. 14, 1858. MANUFACTURED BY REMINGTONS ILION NEW YORK". Number 887 is stamped on the barrel and on the frame under the oil finished, walnut grips.

The arm was patented by Fordyce Beals September 14, 1858, Patent No. 21,478. Prior to 1858 the Remingtons obtained from the inventor the right to manufacture revolvers under his patent rights, and the Beals were the first of the famous Remington made revolvers, all of which were based on Beals patents.

Twelve thousand two hundred and fifty-one Beals revolvers were purchased by the government during the Civil War.

BEALS ARMY PERCUSSION REVOLVER.

Identical in all respects with the Navy Model described above, except that the caliber is .44, the barrel is eight inches long, the total length is thirteen-and-seven-eighths inches, and the weight is 2 pounds, 14 ounces.

BUTTERFIELD ARMY PERCUSSION REVOLVER.

Caliber .41, five shot, single action. Seven inch octagonal barrel rifled with seven grooves. Total length thirteen-and-three-quarters inches. Weight 2 pounds, 10 ounces. The cylinder is one-and-eleven-sixteenths inches long. A brass knife-blade front sight is dovetailed into the barrel. The rear sight is in the frame. The frame and the oval trigger guard are bronze.

A conventional loading lever operates the rammer. A disc primer magazine is located in front of the trigger guard. By removing a thumb screw, a tube of disc primers can be inserted, which are automatically fed to the cones by the action of cocking the hammer. The barrel and cylinder were blued. Shellacked walnut grips. Specimens are known with silver plated frame, trigger guard and grip frame.

The specimen illustrated is marked on the top of the frame "BUTTERFIELD'S PATENT DEC. 11, 1855 PHILADA". Number 431 is marked on the inside of the frame side plates, on the barrel, cylinder, loading lever, hammer-block, trigger and the cylinder lock.

A number of Butterfield revolvers are known *un-marked* except for the serial number, which is in the higher series, about 560 and above. The cause is unknown, but may well have been due to wartime demands, and continuing production in spite of a worn or damaged stamp.

One specimen of a large and heavy Butterfield revolver, probably experimental, is known in caliber .50, with six-and-five-eighths inch barrel and two-and-one-quarter inch cylinder. Total length fourteen-and-one-half inches; weight 5 lbs. 2 oz. It is marked as the army model above, but the parts are not numbered.

The Butterfield mechanical disc primer was patented and manufactured by Jesse Butterfield at Philadelphia, Pa. There is no record of government purchase during the Civil War.

The Butterfield disc primer was also used in the conversion of flintlock muskets to percussion.

Butterfield revolvers are reputed to have been made in calibers .36 and .44. However, no specimen of either was seen in the important collections examined. Possibly these calibers were the results of erroneous measurement of caliber .41, and the larger (.44) may have been the result of calibration of worn specimens. In this connection it is to be noted however, that while the groove diameter of a typical specimen measured .411, the chamber diam- eter of this arm was found to be .44, taking an oversize, soft lead bullet.

COLT MARTIAL PERCUSSION REVOLVER, MODEL 1839. (PATERSON MAKE—WALKER.)

It is believed that in 1839, Samuel Colt, at the request of the Texas Government, produced at his Paterson fac- tory, a martial .44 caliber revolver, which was an im- provement over his previous concealed trigger Patersons, in that it was equipped with a loading lever to operate the rammer, and embodied the additional improvements of a conventional trigger and trigger guard.

Tradition has it that the representatives of the Texas Government, who came East to buy arms for the Lone Star Republic, and possibly Capt. Samuel H. Walker of the Rangers, who may have been one of the agents, if not the sole representative,—contributed suggestions for the improvements incorporated in the Model 1839 revolver. Whether the legend is true or not, these revolvers, of which three hundred or less are reputed to have been made, are referred to as Walker-Colts, in honor of the Ranger captain.

Needless to say, with such a small issue, positive identification is difficult. Legend has it that when Colt received the contract for one thousand revolving pistols in January, 1847, he was unable to find a specimen of the 1839 Model, though but scant seven years had elapsed, and he was forced to design the Model 1847 revolver from memory.

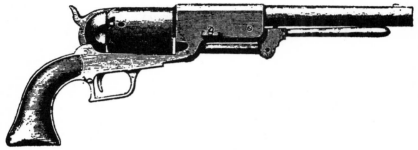

WALKER PISTOL

Cut from "ARMSMEAR"*

of the

Colt Martial Percussion Revolver MODEL 1839

PATERSON Made — WALKER

*NOTE: "Armsmear" a memorial volume to Col. Colt was published in 1865 at the expense and with the collaboration of Mrs. Colt. With reference to the Walker Colt the book states that the original pistol carried by Walker was a part of Colt's collection of arms, leading to the assumption that the above cut was made from that particular arm.

Because of the fact that no specimen of this first Colt martial model was found for many years, many advanced collectors were actually dubious as to such an arm having been made. This disbelief was refuted by Mr. McMurdo Silver, noted Colt collector and student, who supported the existence of this model by statements that the design of the Model 1839 Colt carbines, of which one hundred and sixty were purchased by the government for trial in 1841, was essentially the design of a hand-gun, developed into a carbine, indicating the existence of an 1839 martial revolver. The other point in his argument is the Walker story, and a line drawing which appeared in "Armsmear", a eulogy to Colt, published after his death, for which story and drawing there necessarily was some basis.

In a monograph on early martial Colt revolvers, the owner of a brace of large revolvers, alleged to be the original Walker-Colts, claims that the arms he describes are the only two made of this model, (one of these is marked "Address Col. Colt—London"). The statement is made that law suits within the firm and cessation of manufacturing operations prevented the model from going into production. Time, or discovery of other specimens, may support or confute these claims.

The mere fact that only one specimen may now be known, is not however *per se* proof that a quantity of Paterson made Colt martial revolvers with trigger guard and loading lever, was not produced for the use of the Texas Rangers. Frontier conditions were then turbulent, a state not conducive to preservation of arms. Almost a hundred years have elapsed, and judging from the present relative scarcity of the Model 1848 (Dragoon) revolvers, of which many thousands were made between 1848 and 1860, it is quite possible that positively identified specimens may never be found of the very few hundred which may have been sent to Texas about 1840.

COLT Army Percussion Revolver, MODEL 1847. (Whitneyville Make.)

Illustrated—Fig. 1, Plate 18.

Caliber .44, six shot, single action. Nine inch round barrel semi-octagonal to the rear and rifled with seven grooves. Total length fifteen-and-one-half inches. Weight 4 pounds, 9 ounces. The cylinder is two-and-one-half inches long and has oval cylinder stops. A low knife-blade, brass front sight is set into the barrel; a V-notch rear sight is cut into the hammer lip. Square backed, brass trigger guard, steel back-strap and brass under-strap. The back of the frame at its junction with the grips, is curved.

The cylinder revolves to the right. The rammer is operated by a loading lever hinged on a screw in front of the wedge, which enters the barrel lug from the right side. The loading lever is held in place by a hook-shaped spring, seated under the barrel in front of the pivot screw. The original finish is believed to have been: — blued barrel, cylinder and trigger; case-hardened frame, loading lever and hammer; oil finished, walnut grip. That however is but a surmise, based on traces of factory finish of a specimen examined. The revolvers of this model had all seen hard service and much wear, and on the majority of the known specimens even the engraving can no longer be seen. The usual charge of these holster pistols was fifty grains of black powder, firing a conical bullet.

The revolvers were marked on top of the barrel "AD-DRESS SAML COLT NEW YORK CITY". The bar-rel lug was stamped on the right side over the wedge screw "U.S. 1847". Company designation and number, such as "A COMPANY No. 174", was usually stamped in five places; on the left side of the barrel over the wedge end, on the left side of the frame, in front of the trigger and bolt screws, on the cylinder horizontally from front to rear, on the grip frame, and in front of the trigger guard. The cylinder was engraved with a battle scene of soldiers in action against Indians, though as stated above, all traces of engraving have disappeared from most of the specimens known.

These revolvers were designed by Colt from memory, after his martial Walker Model of 1839, no pattern speci-men being available. Since Colt had no factory of his own at the time, upon receipt of the government contract on Jan. 4, 1847, for one thousand repeating pistols at $28.00 each, Colt arranged to have them manufactured at the Eli Whitney Armory, at Whitneyville, Connecticut. It is reputed however that the barrels and the cylinders for these one thousand Whitneyville Colts were made for Colt on a sub-contract in the Dwight Slate machine shops at Hartford, Connecticut.

COLT ARMY PERCUSSION REVOLVER, MODEL 1848.
(DRAGOON.)
Illustrated—Figs. 2 & 3, Plate 18.

Caliber .44, six shot single action. Seven-and-one-half inch round barrel, semi-octagonal to the rear, and rifled with seven grooves. Total length fourteen inches. Weight

4 pounds, 1 ounce. The cylinder is two-and-three-sixteenths inches long, and has round cylinder stops. A low blade, white brass front sight is set into the barrel; a V-notch rear sight is cut into the hammer lip. The brass trigger guard is square at the rear. Brass back-strap. No bearing wheel on hammer. The revolver is not cut for stock. The back of the frame at its junction with the grips is straight.

The loading lever is hinged on a screw in front of the wedge, and is held to the barrel by a vertical spring-snap catch. The customary finish was charcoal blued barrel, cylinder and trigger; the frame, hammer and loading lever were case-hardened. Shellacked one piece walnut grip.

The specimen illustrated is marked on top of the octagonal part of the barrel "ADDRESS SAML COLT NEW YORK CITY". The left side of the frame is mgarked "COLT'S PATENT" and "US". The cylinder is engraved "MODEL U.S.M.R. COLT PATENT", the initials standing for United States Mounted Rifles, and is also engraved with a scene of soldiers engaged in a running fight with Indians. It also bears the signature of the engraver, "W.L.ORMSBY Sc". (Sc. for sculpist or engraved). Ormsby was the well known New York engraver who cut the original steel roller dies for decorating the cylinders. The revolver is marked on the barrel, loading lever, frame, cylinder, trigger guard and grip frame with number 1362.

This Model 1848 revolver is the next model made by Colt after the completion of the one thousand Model 1847 revolvers which were made for Colt at the Whitneyville Armory. It was manufactured at the Pearl Street

plant equipped by Colt at Hartford, Conn., and is somewhat similar in appearance to the Model 1847, but has
been modified by a shorter barrel and cylinder, reduction
in weight, and an improved loading lever catch.

Model 1848 revolvers were made from 1848 until
about 1860 with a number of minor variations, and from
outward appearance and major details of construction
can be classified into three general types: —

TYPE I. Made in 1848 and 1849. Standard
items were round or slightly oval cylinder stop
slots, and a square back trigger guard. The loading
lever latch projected at the bottom.

Type II. Characterized by improvements consisting of rectangular cylinder slots with inclined
guideways leading to them; a main-spring bearing
wheel to reduce friction and wear, and hammer
rest safety pins between the cones of the cylinder,
fitting into a small recess in the face of the hammer, for safety in carrying the piece loaded. The
loading lever latch operated by side projecting
stud. These improvements were applied to Colt
arms after 1849 and most of them were covered by
Patents of Sept. 10, 1850. Though the majority of
the revolvers of Type 2 were equipped with an
oval trigger guard, specimens equipped with a
square backed trigger guard are quite numerous.
Probabilities are that these earlier type guards
were used until the parts were exhausted, and
were later put on only on special order.

TYPE III. Similar to Type 2, but cut for
shoulder stock and equipped with a two-leaf rear
sight.

In addition to the major variations which permit type classification as above, numerous minor variations are known, such as different barrel lengths, of which the eight inch is the most common; loading lever latches were made in different forms, and revolvers embodying other variations from standard were made up on special order.

The revolver illustrated, with its round cylinder stops and square back trigger guard, belongs to Type 1, of which two thousand were made between 1848 and 1850. It varies from the standard type in that it has hammer safety pins, which may have been put in later, or may have been fitted to a cylinder which had been manufactured before the other changes were embodied.

The revolvers of Type 3, cut for, and fitted with detachable carbine stocks, were adopted by the service from 1858. The stocks had an iron butt-plate, yoke and sling ring; were seventeen-and-five-eighths inches long, and weighed 2 pounds, 8 ounces. A single stock was issued with every two revolvers, and was marked with both their numbers, i.e., 1362, 63. The stocks were made in both solid and canteen models. In the latter the metal canteen was contained in the butt, held by the butt-plate; the mouth emerging at the comb of the stock, capped by a chain fastened screw cap.

In addition to variations in construction there were numerous variations in marking, as well as some omissions of the name of the firm from the barrels. Some cylinders were marked "U.S.M.I." for United States Mounted Infantry; others were marked "U.S.DRAGOONS", whence the name generally applied to this model, and a very few made for navy issue were marked "NAVY".

These arms were also known as Holster Pistols, and Old Model Army Pistols, and were intended to be carried in saddle holsters. The issue was to cavalry and to dragoon regiments of the army. Dragoons were mounted troops, whose combat employment called for fighting on foot, like the infantry, and for whose tactics the pistol-carbine was particularly adapted.

The total output of these revolvers is reputed to have been about 30,000, of which the government purchased about 7,180 at the average price of $24.00 each.

According to a report by Col. H. K. Craig, Chief of Ordnance, before an Ordnance Board convened at the Washington Arsenal in 1858, the government orders for Colt holster pistols were as follows: — (Model 1847 excepted.)

Nov.	2,	1847 —	1,000
Jan.	8,	1849 —	1,000
Feb.	4,	1850 —	1,000
May	8,	1851 —	2,000
May	26,	1853 —	1,000
Jan.	15,	1855 —	1,000
Apr.	23,	1856 —	125
Apr.	29,	1856 —	55

TOTAL　　　　　　7,180 Holster Pistols.

COLT NAVY PERCUSSION REVOLVER, MODEL 1851.
Illustrated—Figs. 1 & 2, Plate 19.

Caliber .36, six shot, single action. Seven-and-one-half inch octagonal barrel rifled with seven grooves. Total length thirteen inches. Weight 2 pounds, 10 ounces. The cylinder is one-and-eleven-sixteenths inches long, and has rectangular notches and safety pins for hammer rest. A small, brass cone front sight is set into the barrel. A V-notch rear sight is cut into the hammer lip. The oval trigger guard and the grip frame are bronze, silver plated.

The hammer of this model is equipped at the base with a small wheel to minimize friction and wear at its contact against the mainspring. Another distinguishing feature of this model was the introduction of "gain twist" rifling in which the relatively slow pitch of the lands in the first two-thirds of the barrel, increased from one turn in about twenty-two inches, to one turn in sixteen inches for the last third of barrel length. The object was to prevent the bullet from jumping the rifling because of the too abrupt pitch in the beginning of bullet travel; then after the bullet had taken the rifling, to insure greater accuracy by the sharper pitch of the lands.

The loading lever is hinged by a screw in front of the wedge. The barrel, cylinder and the trigger were blued; the loading lever, frame and hammer were case-hardened in mottled colors. The walnut grip was shellacked to a high finish. The usual load was a self-consuming combustible cartridge, though loose powder and ball could be used.

The specimen illustrated is marked on the barrel "—ADDRESS COL. SAM∟ COLT NEW YORK U.S. AMERICA—". The frame is stamped on the left side "COLT'S PATENT", and the trigger guard "36 cal." The cylinder is marked "COLT'S PATENT" and "ENGAGED 16 MAY 1843", and is engraved with a scene of a naval engagement. The engraving was to commemorate a battle between the three vessels of the Texas navy, when that state was still the Lone Star Republic, and the Mexican fleet, in which engagement the navy of the young republic, under Colt's friend, Commodore Moore, was victorious over the superior Mexican flotilla. Number 203,231 is stamped on the barrel, wedge, cylinder, trigger guard and grip frame.

Over 200,000 revolvers of this model were made at the Colt Hartford plant between 1851 and 1865.

The early revolvers of this model are known with square-back trigger guards. Other variations have iron back-straps, and a few were made with fluted cylinders supplied on special order. Some of these revolvers were cut for, and supplied with attachable carbine stocks, which in this and succeeding army and navy models differ from the iron mounted M. 1848 Dragoon stocks, in that the *fittings* are *brass*. The stocks were made in both solid and canteen models. In the latter the metal canteen was contained in the butt, held by the butt-plate, the mouth emerging at the comb of the stock. The screw cap was fastened by a small chain to prevent loss. Stock equipped specimens of this model are quite rare. Even rarer are stocks containing coffee mills.

In addition to variations in construction, the markings of these revolvers vary considerably: some specimens are marked "ADDRESS SAML COLT NEW YORK CITY", or "SAML COLT NEW YORK U.S.A." etc., reading in either direction.

COLT PERCUSSION NEW MODEL POLICE PISTOL.
(POCKET NAVY MODEL.)

This arm is supposed to have been issued experiment-ally to the navy for use as an officers side arm, and for that reason is often referred to by collectors as the Pocket Navy Pistol.

The revolver is similar in appearance to the Colt Navy 1851, except that the cylinder is rebated and engraved with a stage-coach hold-up, the frame is lighter, and the .36 caliber barrel is four-and-one-half inches long.

COLT PERCUSSION REVOLVER, MODEL 1855.

These solid frame, side hammer revolvers were patented Dec. 25, 1855, by Elisha K. Root, superintendent of the Colt factory, Pat. No. 13999, the patent being assigned to the Colt Fire Arms Company. The solid frame of the design was to eliminate the weak spot of the open frame Colt. The cylinder pin extracted from the rear and revolved with the cylinder, a design that required the use of a side hammer. The mechanism was rather delicate and unsuitable for military use. Though many small caliber revolvers of this model were manufactured, only a small quantity of the arms of this model are believed to have been made in martial sizes for experimental purposes. Some details of these are given by Mr. Satterlee as follows: —

COLT ARMY PERCUSSION REVOLVER, MODEL 1855.

Caliber .44, six shot, side hammer. Barrel about eight-and-one-eighth inches. Total length fifteen-and-one-eighth inches. Weight 4 pounds, 2½ ounces.

COLT NAVY PERCUSSION REVOLVER, MODEL 1855.

Caliber .36, six shot, side hammer. Barrel seven-and-one-half inches. Total length thirteen-and-three-quarters inches. Weight 3 pounds, 15 ounces.

COLT Army Percussion Revolver, MODEL 1860.
Illustrated—Figs. 3 & 4, Plate 19, and Figs. 1 & 2, Plate 20.

Caliber .44, six shot, single action. Eight inch round barrel rifled with seven grooves. Total length fourteen inches. Weight 2 pounds, 11 ounces. The rebated cylinder is one-and-thirteen-sixteenths inches long and has rectangular cylinder stops and hammer rest safety pins. A low blade, white brass front sight is set into the barrel. A V-notch rear sight is cut into the hammer lip. Oval bronze trigger guard and iron back-strap. The frame and back-strap are notched for a detachable carbine stock.

This model is the first of the Colt "streamline" revolvers, the barrel lug and the loading lever angles having been eliminated and replaced by graceful curves. The loading lever operates the rammer by a ratchet tooth arrangement. Like the 1851 model, the hammer is equipped with a bearing wheel. The usual charge was a self-consuming combustible cartridge, though loose powder and ball could be used. The barrel, cylinder and trigger were blued, the loading lever, frame and hammer case-hardened in mottled colors.

In the specimen illustrated the barrel is marked on top "—ADDRESS COL. SAMl COLT NEW YORK U.S. AMERICA—". The left side of the frame is marked "COLT'S PATENT". The cylinder is stamped "COLT'S PATENT" and "PATENTED SEPT. 10th, 1850", and like the Model 1851 Colt navy, is engraved with the scene of a naval engagement between the Texan and Mexican fleets, and bears the words "ENGAGED 16 MAY 1843". The barrel, wedge, frame, trigger guard and grip frame are numbered 115,234. The oil finished, walnut grip is marked with government inspector's intials "JSD" and "JT" in script in medallions on the left and right side respectively.

This arm was the principal revolver of the Civil War, 107,156 having been furnished to the War Department between Jan. 4, 1861 and Nov. 10, 1863. At the time of issue it was called the New Model Army Pistol.

The extension stock for which the revolver was cut, has a brass butt-plate and a brass yoke which fastens to the revolver by means of a steel clutch. The stock is seventeen inches long and weighs 2 pounds, 5 ounces. The length of the revolver assembled with stock is twenty-six-and-one-half inches. Though the vast majority of the revolvers were cut for extension stocks, the stocks are quite scarce.

This model was made in a number of variations, some were not cut for stock, some had fluted cylinders;—other minor variations were in barrel length, barrel lug, loading lever and weight. A seven-and-one-half inch length was common to the shorter barrel.

COLT NAVY PERCUSSION REVOLVER, MODEL 1861.
Illustrated—Fig. 3, Plate 20.

Caliber .36, six shot, single action. Seven-and-one-half inch round barrel rifled with seven grooves. Total length thirteen inches. Weight 2 pounds, 10 ounces. The cylinder is one-and-eleven-sixteenths inches long, has rectangular cylinder stops and hammer rest safety pins. A low blade, white brass front sight is set into the barrel. The V-notch rear sight is cut into the lip of the hammer. Oval bronze trigger guard and bronze back-strap. The frame is not cut for stock.

The revolver is "streamlined" and similar in appearance to the army Model 1860, except that the caliber is .36, the cylinder is not rebated and the interior of the frame is straight to correspond with the even diameter of the cylinder. The hammer has a bearing wheel, and as in the 1860 model, the loading lever is operated by a ratchet tooth arrangement. The usual charge was a self-consuming cartridge, though loose powder and ball could be used. The barrel, cylinder and trigger were blued; the loading lever, frame and hammer were case-hardened in mottled colors.

In the specimen illustrated the barrel is marked on top "—ADDRESS COL. SAML COLT NEW YORK U.S. AMERICA—". The left side of the frame is marked "COLT'S PATENT" and the trigger guard "36 CAL." The cylinder is stamped "COLT'S PATENT" and like the Model 1851, is engraved with the scene of a naval engagement to commemorate the battle between the Texan and the Mexican fleets. The barrel, wedge, frame, trigger guard and grip frame are numbered 7,112.

The revolver known as the New Model Navy Pistol, was not as popular as the Model 1851 Navy, and relatively few were made. 2,056 were purchased by the War Department between Feb. 17, 1862 and Jan. 20, 1863. The arm was also used in the naval service.

COLT PERCUSSION REVOLVER, MODEL 1862.

Caliber .36, five shot, single action. Six-and-one-half

inch round barrel rifled with seven grooves. Total length eleven-and-one-half inches. Weight 1 pound, 10 ounces. Brass cone front sight; a V-notch rear sight is cut into the hammer lip. The semi-fluted, rebated cylinder is one-and-nineteen-thirty-seconds inches long. Brass, silver-plated trigger guard and back-strap. Walnut grip shellacked to a high finish.

The revolver was marked on the barrel "ADDRESS COL. SAML COLT NEW YORK U. S. AMERICA", and was similar in appearance and "streamline" design to the Colt Model 1861 Navy, but on a smaller frame. The standard finish was blued barrel, and cylinder; case-hardened loading lever, frame and hammer. Though the revolver was called "New Model Pocket Pistol of Navy Caliber" in early Colt literature, it was popular with the officers of the Union Army as a "belt" holster personal side arm, and is often referred to as "Belt Model".

As an arm made for private sale it was manufactured to meet a wide difference in purses and tastes. It was made in a number of finishes, barrel lengths, and minor variations in marking and construction. Some of these revolvers have an iron trigger guard and back-strap, the barrels of others vary in length from four-and-one-half to seven inches, while marking "ADDRESS SAML COLT HARTFORD CONN." is by no means uncommon.

COOPER Navy Percussion Revolver.
Illustrated—Fig. 4, Plate 20.

Caliber .36, five shot, double action. Five-and-seven-eighths inch octagonal barrel rifled with seven grooves. Total length ten-and-three-quarters inches. Weight 1 pound, 12 ounces. The cylinder is one-and-five-eighths inches long. Brass cone front sight; V-notch rear sight is cut into the hammer lip. The back-strap and the oval trigger guard are bronze, silver-plated.

The revolver is similar in appearance to the Model 1851 Colt Navy, except that the cylinder is rebated like the Colt Army 1860. A conventional loading lever oper-ated the rammer. The barrel is dismounted and the cylin-der removed by withdrawing a wedge from the left side. The barrel and cylinder were blued, the frame, loading lever hammer and trigger were case-hardened in mottled colors. The black walnut grips were shellacked to a high finish. The revolver fired a combustible cartridge or could be loaded with loose powder and ball.

The specimen illustrated is marked on the barrel "COOPER FIREARMS MFG. CO. FRANKFORD PHILA PA. PAT. JAN. 7 1851, APR. 25 1854, SEP. 4 1860, SEP. 1 1863, SEP. 22 1863". Number 5641 is stamped on the barrel, cylinder, loading lever and grip frame.

The revolver was patented by James Maslin Cooper of Pittsburg, Penna., Patent Nos. 29,864 and 40,021, and was manufactured by the Cooper Firearms Co., at Frank-ford, Pa. Though resembling a Colt in outline, it was not intended as an imitation. This double action revolver was well made and finished, but the design of the frame, while well adapted for single action, made its use difficult as a double action weapon. However the arm might have been more popular, had the firm had facilities for quantity pro-duction. The Cooper Arms Company went out of busi-ness in 1869.

FREEMAN Army Percussion Revolver.
Illustrated—Fig. 5, Plate 20.

Caliber .44 six shot, single action. Seven-and-one-half inch round barrel rifled with six grooves. Total length twelve-and-one-half inches. Weight 2 pounds, 13 ounces. The cylinder is one-and-seven-eighths inches long. Steel blade front sight. The rear sight is in the frame.

The cylinder and the two-part cylinder pin may be removed by moving forward a slide on the right side of the frame, in front of the cylinder. The rammer is operated by a conventional loading lever. The frame, barrel and cylinder are blued, the hammer and loading lever are case-hardened. Walnut grips, oil finished. The revolver fired a self-consuming combustible cartridge, or could be loaded with loose powder and ball.

The specimen illustrated is marked on the frame over the cylinder, "FREEMAN'S PAT. Decr. 9, 1862, HOARD'S ARMORY, WATERTOWN, N. Y." Number 1342 is stamped on the barrel, frame, loading lever, cylinder and cylinder pin.

The revolver was patented by Austin T. Freeman of Binghamton, N. Y., December 9th, 1862, Patent No. 37,091 and was manufactured by C. B. Hoard at Watertown, N. Y. The Rogers & Spencer revolver is a development of the Freeman, the patent having been purchased by that firm from the inventor.

The Freeman revolver is one of the rarer weapons of the Civil War. There is no record of government purchase, though Hoard's Armory made 12,800 muskets on contract during the Civil War.

JOSLYN ARMY PERCUSSION REVOLVER.
Illustrated—Fig. 1, Plate 21.

Caliber .44, five shot, single action. Eight inch octagonal barrel rifled with five grooves. Total length fourteen-and-three-eighths inches. Weight 3 pounds. The cylinder is two-and-three-sixteenths inches long. A steel, low, knife-blade front sight is dovetailed into the barrel. The rear sight is in the frame. The oil-finished walnut stocks are coarsely cross checked between the upper and lower grip screws.

A curved side-hammer on the right of the frame strikes through the center of the frame. The cylinder is removed by releasing a screw in the frame behind the hammer and withdrawing the cylinder pin from the rear. The cylinder is rotated by a stud in a disc in the rear of the frame, operated by the hammer. A conventional loading lever operates the rammer. All metal parts were blued. The revolver fired a self-consuming, combustible cartridge, or could be loaded with loose powder and ball.

The specimen illustrated is marked on top of the barrel "B. F. JOSLYN PATd MAY 4th 1858". Number 1628 is stamped on the barrel, loading lever, trigger guard, cylinder, cylinder pin and sleeve and on the frame.

The revolver was patented by Benjamin F. Joslyn of Stonington, Conn. Pat. No. 20,160, and was manufactured by the Joslyn Firearms Company of the same place.

Joslyn Army revolvers are also found marked "B. F. JOSLYN, STONINGTON, CONN." and "B. F. JOSLYN WORCESTER MASS." with a slight difference of a fraction of an inch in barrel and total length. The Worcester arm was made by W. C. Freeman under Joslyn patents at Worcester, Mass., on a government contract. A total of 1100 Joslyn revolvers were purchased by the government during the Civil War for army and navy use. Of these 875 were purchased by the army in 1862.

LEAVITT Percussion Revolver.

Caliber .40, six shot, single action, hand turned cylinder. Six-and-three-quarters inch round barrel. Total length thirteen-and-seven-eighths inches. Walnut grip.

The specimen described is marked "LEAVITT'S PATENT MANUFACTURED BY WESSON STEPHENS & MILLER HARTFORD CT."

The revolver was manufactured between 1837 and 1839 under Patent No. 182, granted Daniel Leavitt April 29, 1839, which described the arm as a "powder and ball single action solid frame revolver with tip-up action released by pressing a lever inside the front end of the trigger guard."

Leavitt had the arms manufactured in the shop of Edwin Wesson, rifle maker of Hartford, at which time began Leavitt's association with Wesson which eventually resulted in an improved arm manufactured by the Massachusetts Arm Company, the changes incorporating Wesson's patent of mechanical revolution of the cylinder by means of bevel gears. (See Wesson & Leavitt, early and later types.)

MANHATTAN NAVY PERCUSSION REVOLVER.
Illustrated—Fig. 2, Plate 21.

Caliber .36, five shot, single action. Six-and-one-half inch octagonal barrel rifled with five grooves. Total length eleven-and-one-half inches. Weight 2 pounds. The cylinder is one-and-nine-sixteenths inches long. A white brass, knife-blade front sight is set into the barrel. A V-notch rear sight is cut into the lip of the hammer. The backstrap and the oval trigger guard are of bronze, silverplated. The walnut stocks are shellacked to a high finish.

The construction and design are similar to the Colt Model 1851 Navy, which this revolver closely imitated, but with the addition of a spring plate interposed between the caps and the hammer. The barrel, cylinder and trigger are blued, the loading lever, frame and hammer are case-hardened. The revolver fired a combustible cartridge or loose powder and ball.

The specimen illustrated is marked on the top of the barrel "MANHATTAN FIRE ARMS CO. NEWARK N. J. PATENTED MARCH 8, 1864". The cylinder is marked "PATENTED DEC. 27, 1859", and is engraved with five scenes in medallions. The barrel, frame, wedge, cylinder, cylinder shaft, trigger guard and butt-frame are numbered 53,356.

On Dec. 27, 1859, Patent No. 26,641 was issued to J. Gruler and A. Rebetey for a single action revolver. It is uncertain whether the Manhattan revolver was manufactured under this patent. The spring-plate, whose function was to deflect the possible back-flash from the cap, was patented March 3, 1864 by Ben Kitteredge of Cincinnati, Ohio, Pat. No. 41,848. There is no record of government purchase during the Civil War.

METROPOLITAN NAVY PERCUSSION REVOLVER.
Illustrated—Fig. 3, Plate 21.

Caliber .36, six shot, single action. Seven-and-one-half inch octagonal barrel rifled with seven grooves. Total length thirteen inches. Weight 2 pounds, 8½ ounces. The cylinder is one-and-eleven-sixteenths inches long. Brass cone front sight; a V-notch rear sight is cut into the hammer lip. The oval trigger guard and the back-strap are brass, silver-plated.

The revolver is a close imitation of the Model 1851 Colt Navy, in design and construction. Later models of this arm were made with hammer rest notches in the cylinder, in the shoulders between the cones. The barrel, frame, cylinder and trigger were blued, the hammer and loading lever were case-hardened. The walnut grip was shellacked to a high finish. The arm fired the usual combustible cartridge of the time, or could be loaded with loose powder and ball.

The barrel of the specimen illustrated is marked "METROPOLITAN ARMS CO. NEW YORK". The cylinder is engraved with a scene of an engagement between the Union fleet and Confederate land batteries, and is inscribed "NEW ORLEANS APRIL 1862 W. L. ORMSBY Sc". The barrel, loading lever, frame, cylinder, trigger guard and grip frame bear the number 4810.

The revolver was made by the Metropolitan Arms Company, which was established in 1859, and engaged in the manufacture of revolvers in imitation of Colts, after the expiration of the Colt patents.

METROPOLITAN Percussion Revolver, (Pocket Navy.)
Illustrated—Fig. 4, Plate 21.

Caliber .36, five shot, single action. Five-and-fifteen-thirty-seconds inch round barrel rifled with seven grooves. Total length ten-and-five-eighths inches. Weight 1 pound, 10 ounces. The semi-fluted, rebated cylinder is one-and-nine-sixteenths inches long, and has hammer rest notches in the shoulders between the cones. Brass cone front sight; the rear sight is cut into the hammer lip. The oval brass trigger guard and the back-strap are silver-plated. Blued barrel and cylinder; case-hardened loading lever, frame, hammer and trigger. The walnut grip is shellacked to a high finish.

This Metropolitan revolver of streamline design was made in imitation of the Colt Model 1862, so called "Pocket Navy" percussion revolver. The arm used a self-consuming, combustible cartridge, or could be loaded with loose powder and ball.

The specimen illustrated is marked on top of the barrel "METROPOLITAN ARMS CO. NEW YORK". Number 1886 is stamped on the barrel lug, loading lever, wedge, cylinder, frame, trigger guard and grip frame. These revolvers were also made in nickel finish.

PETTINGILL Army Percussion Revolver.
Illustrated—Fig. 5, Plate 21.

Caliber .44, six shot, double action, hammerless. Seven-and-one-half inch octagonal barrel rifled with six grooves. Total length fourteen inches. Weight 3 pounds. The cylinder is two-and-one-quarter inches long. A brass cone front sight is set into the barrel. The rear sight is in the frame. Oil finished walnut grips.

The hammer is concealed within the solid frame. Pressure on the trigger cocks the hammer, revolves the cylinder and fires the piece. A removable set screw on the left side of the frame permits the withdrawal of a conventional loading lever, the attached cylinder pin, and the removal of the cylinder. Blued barrel and blued or browned frame. Case-hardened loading lever. The revolver used a self-consuming, combustible cartridge, or could be loaded with loose powder and ball.

The specimen illustrated is marked on the top of the frame "PETTINGILL'S PATENT 1856", and "RAY-MOND & ROBITAILLE PATENTED 1858". Number 3008 is stamped in the trigger guard and on the grip frame. Inspector's initials "NW" are stamped in script in a medallion on the left grip.

The mechanism is the invention of C. S. Pettingill of New Haven, Conn., Pat. No. 15,388, July 22, 1856, and of Edward A. Raymond and Charles Robitaille of Brooklyn, N. Y., co-patentees, Pat. No. 21,054, July 27, 1858. The revolver was manufactured by Rogers & Spencer of Willowdale, N. Y. 2,001 were delivered to the government on contract during the Civil War. The arm did not prove serviceable, as the mechanism was complicated and delicate, and could not stand the gaff under field service conditions.

PETTINGILL NAVY PERCUSSION REVOLVER.
Illustrated—Fig. 6, Plate 21.

Similar in construction and appearance to the Army Model above, except that the caliber is .36, the barrel is four-and-five-eighths inches. Total length ten-and-one-half inches. Weight 1½ pounds.

PLANT ARMY REVOLVER, PERCUSSION OR CUP-PRIMER CARTRIDGE.

This arm, though equipped with an interchangeable percussion cylinder, was manufactured primarily as a cup-primer cartridge revolver, and is described in the chapter on cartridge revolvers.

REMINGTON ARMY PERCUSSION REVOLVER, MODEL 1861.
Illustrated—Fig. 1, Plate 22.

Caliber .44, six shot, single action. Eight inch octagonal barrel rifled with five grooves. Total length thirteen-and-three-quarters inches. Weight 2 pounds, 14 ounces. The cylinder is two inches long. On the barrel is mounted a German silver, cone front sight. The rear sight is in the frame. Oval brass trigger guard. A conventional loading lever operates the rammer. Oil finished walnut stocks.

The loading lever of this model is cut out at the top so that the cylinder pin can be slid forward and the cylinder removed without lowering the lever. Blue finish except the hammer which was case-hardened. The usual charge was a self-consuming, combustible cartridge, though loose powder and ball could also be used.

The specimen illustrated is marked on the top of the barrel "PATENTED DEC. 17, 1861 MANUFACTURED BY REMINGTON'S ILION N. Y."

This revolver was made by Remingtons under patent issued to Wm. Elliott of Plattsburg, N. Y., for a single action revolver, Pat. No. 33,932, dated Dec. 17, 1861.

REMINGTON Navy Percussion Revolver, MODEL 1861.

Identical in all respects with the Army Model 1861 described above, except that the caliber is .36, the barrel is seven-and-three-eighths inches long, the total length is thirteen-and-one-eighth inches, and the weight is 2 pounds, 8 ounces.

REMINGTON Army Percussion Revolver, NEW MODEL. Illustrated—Figs. 2, 3 & 4, Plate 22.

Caliber .44, six shot, single action. Eight inch octagonal barrel rifled with five grooves. Total length thirteen-and-three-quarters inches. Weight 2 pounds, 14 ounces. The cylinder has intermediate hammer rest recesses and is two inches long. The barrel carries a blade front sight with a rounded base. The rear sight is in the frame. Oval brass trigger guard. Oil finished walnut grips. The New Model Remington may be easily recognized by the threads of the barrel which are visible where the barrel adjoins the cylinder.

Like the Beals revolver, the cylinder of the New Model is removed by lowering the conventional loading lever and withdrawing the cylinder pin to the front. Blue finish except for the hammer which is case-hardened. The revolver fired a self-consuming, combustible cartridge, or could be loaded with loose powder and ball.

The specimen illustrated is marked on the barrel "PATENTED SEPT. 14, 1858. E. REMINGTON & SON, ILION, NEW YORK, U.S.A." and "NEW MODEL". The left grip is stamped with inspector's initials "O.W.A." in script in a rectangle.

In this New Model the Remingtons overcame the slightly faulty design of the earlier Beals and of Model 1861, and provided a splendid weapon for its day. Its solid frame, easily removable cylinder and rugged construction, made it a dependable military weapon, and next to the Colt, the best known of the various Civil War types. 125,314 Remington revolvers were purchased by the U. S. government during the Civil War.

REMINGTON Navy Percussion Revolver, NEW MODEL.

Identical in all respects with the Army Model described above, except that the caliber is .36, the barrel is seven-and-three-eighths inches long, the total length is thirteen-and-three-eighths inches. Weight 2 pounds, 8 ounces.

REMINGTON-RIDER Navy Percussion Revolver.
Illustrated—Fig. 5, Plate 22.

Caliber .36, six shot, double action. Six-and-one-half inch octagonal barrel rifled with five grooves. Total length eleven-and-one-half inches. Weight 2 pounds, 1 ounce. The full fluted cylinder is one-and-seven-eighths inches long. Brass cone front sight; rear sight is a V-notch cut in the frame. Oval brass trigger guard. Blued barrel, frame and cylinder; case-hardened hammer. Walnut grips shellacked to a high finish.

The specimen illustrated was marked on the top of the barrel "MANUFACTURED BY REMINGTON'S ILION N. Y. RIDER'S PT. AUG. 17, 1858, MAY 3, 1859".

The revolver was made from the designs of Joseph Rider, Remington's factory superintendent, who patented its double action feature on May 3, 1859, Patent No. 23,861.

ROGERS & SPENCER ARMY PERCUSSION REVOLVER.
Illustrated—Fig. 6, Plate 22.

Caliber .44, six shot, single action. Seven-and-one-half inch octagonal barrel rifled with five grooves. Total length thirteen-and-three-eighths inches. Weight 3 pounds, 2 ounces. The cylinder is two inches long. Brass cone front sight. Rear sight in frame.

A conventional loading lever operates the rammer. The cylinder may be removed by unscrewing a double retaining screw in the front of the frame, and withdrawing the loading lever, with the attached cylinder pin, to the front. The barrel, frame and cylinder were blued; the loading lever and hammer case-hardened in mottled colors; the trigger was burnished bright. The square bottom, bell-shaped, black walnut grips were oil finished. The revolver fired a self-consuming, combustible cartridge, or could be loaded with loose powder and ball.

The specimen illustrated is marked on top of the frame "ROGERS & SPENCER UTICA N. Y." Number 2199 is stamped on the frame, cylinder, and grip frame. Government inspector's initials "RPB" in script in a rectangle are stamped on the left grip.

The revolver is a development of the Freeman army revolver and was made by Rogers & Spencer at Willowdale, N. Y., about seven miles south of Utica. Rogers & Spencer improved and refined the arm made under the

purchased Freeman patents, and their revolver was a handsome, sturdy, well balanced and fine-handling weapon.
5,000 Rogers & Spencer revolvers were obtained by the
War Department between Jan. 30th and Sept. 26th, 1865,
too late for use in the Civil War. This accounts for the
fact that so many of them are found in fine condition.

SAVAGE-NORTH Navy Percussion Revolver.
Illustrated—Fig. 1, Plate 23.

Caliber .36, six shot, single action. Seven-and-three-
sixteenths inch octagonal barrel rifled with five grooves.
Total length fourteen inches. Weight 3 pounds, 7 ounces.
The cylinder is two inches long. Brass cone front sight;
the V-notch rear sight is dovetailed into the frame under
the hammer loop. The arm has no trigger guard. Oil
finished walnut grip.

The revolver has a bronze frame with a spur on the
back-strap. A figure-8 trigger cocks the hammer and operates the cylinder. The rearward movement of the ring
trigger also draws the cylinder backwards, away from the
barrel, before the cylinder starts to revolve. Releasing the
ring trigger moves the cylinder mechanically forward.
The end of each cylinder chamber is chamfered, and when
the cylinder moves forward into firing position, the bevelled breech of the barrel fits against the chamber, forming
a nearly gas-tight joint. The piece is then fired with the
regular trigger, which fits into the upper space of the
figure-8 trigger. The hammer is slightly offset in the
frame. A conventional loading lever operates the rammer. The barrel and cylinder were blued, the loading lever

and hammer were case-hardened. The revolver fired a load of loose powder and ball, or a combustible cartridge.

This model was made in a number of minor variations, and with iron as well as bronze frames.

An interesting feature of the Savage-North arm is the provision made for adjusting the fit between the barrel and the cylinder; a means for securing as nearly a gas-tight joint as is possible in a revolver. The detachable plate in rear of the cylinder contains an adjusting device consisting of a nut for moving the short bearing pin forward or to the rear as desired. The front of a lock-toggle bears against and supports this bearing pin, and as the bearing pin is moved to the rear by the action of the nut, the cylinder is forced forward to fit tightly against the breech of the barrel. (See sketch.)

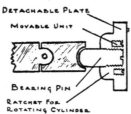

SAVAGE-NORTH

The specimen illustrated is marked on the barrel "E. SAVAGE MIDDLETOWN CT. H. S. NORTH PATENTED JUNE 17, 1856". Number 17 is marked on the barrel, loading lever, hammer, trigger guard and ring trigger.

The mechanism was invented by Henry S. North, and was patented June 17, 1856, Patent No. 14,144. The revolver was manufactured by Edward Savage who made arms at Middletown, Conn., between 1856 and 1859.

In 1860 the revolver was improved by Savage and North, Patent No. 28,331. The joint patentees organized the Savage Repeating Firearms Corporation, which in addition to the second model Savage revolver, made 25,520 rifled muskets for the government during the Civil War.

SAVAGE Navy Percussion Revolver.
Illustrated—Fig. 2, Plate 23.

Caliber .36, six shot, single action. Seven-and-one-eighth inch octagonal barrel rifled with five grooves. Total length fourteen-and-one-quarter inches. Weight 3 pounds, 6 ounces. The cylinder is two-and-three-sixteenths inches long. Brass cone front sight. The V-notch rear sight with a round base is set into the frame under the hammer loop. Oil finished walnut grips.

The hammer is slightly offset in the frame. An enormous trigger guard contains two triggers of which the front, (upper), drops the hammer and fires the piece, while the lower, a ring trigger, cocks the hammer and revolves the cylinder. The rearward movement of the ring trigger also draws the cylinder backwards, away from the bevel shaped rear of the barrel, before the cylinder starts to revolve. Releasing the ring trigger causes the cylinder to slide forward forming a gas-tight joint with the chamber. The piece is then fired in the regular manner with the forward, (upper) trigger. A conventional loading lever operates the rammer. The barrel, frame and cylinder were blued. The loading lever, hammer, triggers and the trigger guard were case-hardened in mottled colors. The revolver used a self-consuming, combustible cartridge, or could be loaded with loose powder and ball.

In the specimen illustrated the barrel is marked "SAVAGE R.F.A. CO. MIDDLETOWN CT. H. S. NORTH PATENTED JUNE 17, 1856. JAN. 18, 1859, MAY 15, 1860". Inspector's initials "WCT" and "IT" in script in medallions are stamped on the right and left grip respectively.

The mechanism is the invention of Hy. S. North and Edward Savage of Middletown, Conn. Patent Number 28,331. 11,284 Savage Navy revolvers were purchased by the government during the Civil War.

STARR ARMY PERCUSSION REVOLVER.
Illustrated—Figs. 3 & 4, Plate 23.

Caliber .44, six shot, double action. Six inch round barrel rifled with six grooves. Total length eleven-and-five-eighths inches. Weight 2 pounds, 15 ounces. The cylinder is one-and-seven-eighths inches long. A steel blade front sight is dovetailed into the barrel. A V-notch rear sight is cut into the hammer lip. (Fig. 4)

The revolver is provided with a removable screw on the right side of the frame which permits the barrel to drop and the cylinder to be removed. A conventional loading lever operates the rammer. The barrel, frame and cylinder were blued; the hammer, loading lever and trigger were case-hardened in mottled colors. The revolver fired a self-consuming, combustible cartridge or could be loaded with loose powder and ball.

The specimen illustrated is marked on the right side of the frame "STARR'S PATENT JAN. 15, 1856" and on the left side "STARR ARMS CO. NEW YORK".

Number 6229 is stamped on the cylinder and on the frame. The wood grips are stamped with inspector's initials "ADK" in script in a rectangle on both sides.

The revolver was the invention of Eben T. Starr of New York City, Pat. No. 14,118, Jan. 15, 1856, and Pat. No. 30,843, Dec. 4, 1860, for a double action top break revolver. Starr revolvers were manufactured by the Starr Arms Company at Yonkers and at Binghamton, N. Y.

This army revolver was also made in *single action,* with an *eight inch barrel.* Total length thirteen-and-three-quarters inches. Weight 3 pounds, 1 ounce. (Fig. 3).

47,952 Starr revolvers were purchased by the government during the Civil War. The vast majority were of the army type. The Navy, cal. .36 Models are comparatively scarce.

The Starr plant located in Binghamton was later sold to "Jones of Binghamton — He pays the freight", who made scales for a good many years. He was General Edward F. Jones, who commanded the Massachusetts regiment that was fired on while going through Baltimore early in the Civil War. The Binghamton street leading to the former location of the plant, is still called Starr Avenue.

STARR NAVY PERCUSSION REVOLVER.
Illustrated—Fig. 5, Plate 23.

This double action arm is similar in all respects to the double action army model with the six inch barrel, except that the caliber is .36, the cylinder is two-and-one-quarter inches long, the weight is 3 pounds, 3 ounces, and the total length is twelve inches, due to a different bend of the grip frame.

UNION NAVY PERCUSSION REVOLVER.

Caliber .36, six shot, single action. Seven-and-seven-eighths inch octagonal barrel rifled with six grooves. Total length thirteen-and-one-half inches. Weight 2 pounds, 8 ounces. The cylinder is one-and-three-quarters inches long. Brass cone front sight; the V-notch rear sight is in the frame. Oval brass trigger guard. The barrel, frame, cylinder, loading lever and hammer were blued. Varnished walnut grips.

The revolver is equipped with a conventional loading lever, and closely resembles the Whitney Navy revolver. The usual charge was a self-consuming, combustible cartridge, though loose powder and ball could also be used.

The specimen described is marked on the barrel "UNION ARMS CO." Number 430 is stamped on the barrel, loading lever, cylinder and frame.

The Union Arms Company of New York, N. Y., and Newark, N. J., received a contract for 25,000 Springfield muskets on November 14, 1861, at $20.00 each, to be delivered at the rate of two thousand monthly, after five months from the date of contract. It is not believed that the contract was fulfilled.

Unions Arms Company is also listed in the Hartford, Conn., City Directory at No. 2 Central Row, in 1861.

WALCH NAVY PERCUSSION REVOLVER.
Illustrated—Figs. 1 & 2, Plate 24.

Caliber .36, twelve shot, single action. Six inch octagonal barrel rifled with six grooves. Total length twelve-and-one-quarter inches. Weight 2 pounds, 4 ounces. The cylinder is one-and-seven-eighths inches long. Brass front sight; no rear sight. Blued barrel, frame and cylinder; case-hardened hammers and triggers. Shellacked walnut grips, checked at the lower half.

The revolver is equipped with two hammers, two triggers and twelve cones, and was loaded with two loads, one on top of the other, in each of the six cylinder chambers, each load being fired by its own cone. A conventional loading lever was used to operate the rammer. The arm used self-consuming, combustible cartridges.

The first of the two specimens illustrated is handsomely engraved all over, is unmarked, and probably was a presentation piece. The usual marking of these arms was "WALCH FIREARMS CO. NEW YORK PAT. FEB. 8, 1859".

The arm was invented and patented by J. Walch who obtained Patent No. 22,905 on Feb. 8, 1859, for a revolver with long chambers, holding two loads each, fired by separate cones.

WARNER NAVY PERCUSSION REVOLVER.
Illustrated—Fig. 3, Plate 24.

Caliber .36, six shot, single action. Six inch round barrel rifled with seven grooves. Total length twelve-and-one-half inches. Weight 2 pounds, 2 ounces. A brass cone front sight is set into the barrel; a V-notch rear sight is on the barrel extension. Iron trigger guard and frame. Walnut grips. Blued finish.

The revolver illustrated is equipped with a center hammer. A conventional loading lever operates the rammer. The usual load was loose powder and ball, or a combustible cartridge.

The arm is marked on the frame "WARNER'S PATENT JAN. 7, 1851", and on the top of the barrel "SPRINGFIELD ARMS COMPANY."

The revolver was invented by James Warner, manager of the Springfield Arms Company, and was manufactured by that concern under patents of Jan. 7, 1851, Patent No. 7,894, and July 15, 1851, Patent No. 8,229.

James Warner, believed to have been a former employee of the Massachusetts Arms Co., obtained three patents for revolver improvements, and manufactured three types of revolvers under the above patents and Patent No. 17,904, granted July 28, 1857.

The first model resembled the Leavitt revolver in outline, as the cylinder shaft formed the lower part of the frame. The arm had a side hammer similar to the Leavitt, and two triggers, of which the front one turned the cylinder, and the rear one released the hammer.

In the second model the releasing trigger was reduced to a latch, which was tripped by pressure on the front trigger.

In the third model is found the conventional center hammer, single trigger design described above.

WARNER Army Percussion Revolver.

Similar in design to the navy model described above. Caliber .44, six inch barrel. The cones are covered by a shield. A heavy weapon.

James Warner & Co. pistol factory is listed in the Springfield City Directory from 1850 until 1869.

WESSON & LEAVITT ARMY PERCUSSION REVOLVER.
Illustrated—Fig. 4, Plate 24.

Caliber .40, six shot, single action. Seven-and-one-eighth inch round barrel rifled with seven grooves. Total length fifteen inches. Weight 4 pounds, 6 ounces. The cylinder is two-and-one-quarter inches long. A brass blade front sight and a small V-notch, open rear sight are set on the barrel, and the barrel extension respectively. Oval brass trigger guard. Walnut grip.

The mechanically turned cylinder is revolved by cocking the hammer, through the application of the Edwin Wesson Patent No. 6669 for revolving the cylinder by use of bevel gears, issued to Wesson August 28, 1849. Turning a catch in front of the cylinder around the barrel, permits the barrel to be raised and the cylinder to be removed. The finish of the specimen illustrated was:— blued barrel and back-strap; the cylinder, frame and hammer, case-hardened, or plated to a smooth, gray color.

The barrel extension is marked on the top "MASS. ARMS CO. CHICOPEE FALLS"; the frame is stamped "WESSON'S & LEAVITT'S PATENT", and the barrel locking device is marked "PATENTED NOV. 26, 1850".

These revolvers were made from 1849 until 1851, when their manufacture was discontinued as the result of the loss of a patent infringement suit.

The Massachusetts Arms Company was organized by the heirs and kin of Edwin Wesson, for the manufacture of revolvers under the Wessons patents. Edwin Wesson, who died in 1850, had been previously associated with Daniel Leavitt in the manufacture of the LEAVITT revolver, made with a hand turned cylinder under the Leavitt Patent No. 182, April 29, 1837. At the time of his death Wesson had a patent pending for an improvement embodying mechanical operation, the patent rights to which formed a part of the Wesson inheritance.

With the receipt of the patent right to mechanical operation, Patent No. 6669, dated Aug. 28, 1849, the Massachusetts Arms Company, which had been making the older hand-turned models, started the production of the new model revolver, which in addition to mechanical operation had several other desirable features, among them a frame that pivoted forward of the hammer, permitting easy and rapid removal of the cylinder. The new models were barely on the market when the company was faced with a suit brought by Colt Patent Fire Arms Company for the infringement of Colt patents. The Colt Company who was represented by Edward S. Dickinson, foremost patent attorney of the day, won the suit, though the Massachusetts Arms Co. had retained Hon. Rufus Choate, one of the famous lawyers of the era, as its counsel. The Massachusetts Arms Company had to cease the manufacture of revolvers under the Edwin Wesson patents until the expiration of the Colt patents for a mechanically operated cylinder in the fall of 1856.

WESSON & LEAVITT Navy Percussion Revolver.

Similar in design and construction to the army model above, except that the caliber was .36, and the seven inch barrel was tinned.

WESSON & LEAVITT Percussion Revolver.

Illustrated—Fig. 5, Plate 24.

Caliber .40, six shot, single action, mechanically turned cylinder. Six-and-one-quarter inch round barrel rifled with seven grooves. Total length thirteen-and-three-quarters inches. Weight 3 pounds, 10 ounces. The cylinder is two-and-one-quarter inches long. Brass blade front sight; there is no rear sight. Oval brass trigger guard, walnut grips, blued finish.

The cylinder may be removed by turning a locking catch on the barrel, tilting the barrel up and sliding the cylinder off the shaft. This model embodies the Edwin Wesson improvement of revolving the cylinder by cocking the hammer through the use of bevel gears, Patent No. 6669, issued August 28, 1849. The appearance of these revolvers on the market brought on a suit for patent infringement by Colt Patent Fire Arms Co. The Massachusetts Arms Company lost the case, (see Massachusetts Arms Co. above), and the manufacture of these revolvers was discontinued.

The specimen illustrated is marked on the side of the lock frame "WESSON'S & LEAVITT'S PATENT", and on the top of the barrel extension "MASS. ARMS CO. CHICOPEE FALLS". The cylinder is etched with a floral and leaf design. The hammer is engraved.

This arm is a development of the Leavitt revolver, which had been manufactured in the Wesson shops under the Leavitt patents of 1837. In the earlier model of this revolver the cylinder was turned by hand for each shot.

WHITNEY NAVY PERCUSSION REVOLVER.
Illustrated—Fig. 6, Plate 24.

Caliber .36, six shot, single action. Seven-and-five-eighths inch octagonal barrel rifled with seven grooves. Total length thirteen-and-one-eighth inches. Weight 2 pounds, 9 ounces. The cylinder is one-and-three-quarters inches long. A small brass cone front sight is set into the barrel. A V-notch rear sight is in the frame and a U-shaped notch is cut into the hammer lip to clear the line of sight. Bronze trigger guard. Oil finished walnut grips.

The revolver is similar in mechanism to the New Model Remington. The cylinder may be removed by withdrawing a screw-held lug on the left side of the frame, and the loading lever and the attached cylinder pin can be pulled out to the front. The barrel, frame and cylinder were blued. The loading lever and hammer were case-hardened. The revolver used a self-consuming, combustible cartridge, or could be loaded with loose powder and ball.

The specimen illustrated is marked on top of the barrel, "E. WHITNEY NEW HAVEN". The cylinder is marked "WHITNEYVILLE", and is engraved with a coat of arms and a naval scene. Number 23,379 is stamped on the loading lever, barrel and cylinder.

The revolver was manufactured at the Whitneyville Armory, near New Haven, Connecticut, by Eli Whitney, son of the famous inventor of the cotton gin, who also was one of the early American arms contractors. Upon the death of Eli Whitney, Sr., in 1825, the Whitney Armory was managed by trustees until young Whitney's coming of age in 1842. The superintendent of the Whitney plant at about that time was Thomas Warner, who had been the master armorer of the Springfield Armory and was let out in 1841, when the government abolished civilian superintendents. The revolver was probably manufactured under the Whitney revolver patent No. 11,447, of Aug. 1, 1854.

In addition to the manufacture of the Whitney Navy revolver of which 11,214 were purchased by the government during the Civil War, Whitney had had an earlier contract for Model 1842 (often referred to as Model 1841) percussion rifles, some of which were shipped by boat to New Orleans and were issued to the 1st Mississippi Regiment, commanded by Col. Jefferson Davis in 1847.

During the Civil War, Whitney had contracted for 40,000 Springfield rifle muskets Dec. 24, 1861, and for 15,000 Oct. 17, 1863. The Whitney Armory ceased operations in 1888.

Chapter 2.

MARTIAL CARTRIDGE REVOLVERS

NOTES ON CARTRIDGE REVOLVERS — CONVERSIONS

CARTRIDGE REVOLVERS —

BACON —

COLT MODEL 1872 —

COLT MODEL 1878 —

COLT MODELS 1892-94-96, ARMY —

COLT MODELS 1901-03 ARMY —

COLT MODELS 1889-95 NAVY

COLT MODEL 1907 —

COLT MODEL 1909 —

COLT MODEL 1917 —

FOREHAND & WADSWORTH —

HOPKINS & ALLEN —

MERWIN & HULBERT —

PLANT —

POND —

PRESCOTT —

REMINGTON MODEL 1874 —

SMITH & WESSON No. 2 —

SMITH & WESSON MODEL 1869 —

SMITH & WESSON MODEL 1875 —

SMITH & WESSON MODEL 1881 —

SMITH & WESSON MODEL 1899 —

SMITH & WESSON MODEL 1917 —

MARTIAL CARTRIDGE REVOLVERS.

U. S. MARTIAL CARTRIDGE REVOLVERS.

After the close of the Civil War, the wide introduction of the metallic cartridge marked the end of the
percussion era.   Though officially the Smith & Wesson
Company enjoyed the monopoly of the practical metallic
cartridge revolver until the expiration of the Rolin White
patents in 1869, many percussion arms, principally Colts,
were converted to cartridge type.   Some of the more
common alterations consisted of cutting away the part of
the outer end of the cylinder, containing the cones or nipples, and replacing that section by a disc of cylinder diameter containing a loading gate.   This was the usual factory
conversion, and was completed by adding a spring rod
ejector to the barrel.   Another method used by gunsmiths
was to braze or bolt a section cut from an extra cylinder
to the cut-away section of the original cylinder on which
the ratchet was left intact, and to lengthen the hammer
nose to strike a .44 Henry rim fire, or later a center fire
cartridge.

Colt army and navy models were also factory converted to take a brass, center fire, rimless cartridge, tapering from front to rear, and loading from the front of the
cylinder by using the rammer, the cartridge being held in
place by forcing the oversize lead bullet.   The system was
patented by Thuer, Sept. 15, 1868, and adapted by Colt;
but judging from the relatively few conversions made on
this system, the method was not a success.

With the expiration of the Rolin White patents, the modern gas tight cartridge brought about the universal adoption of the breech loading cartridge system to revolvers.

Towards the end of the century the development of progressive burning smokeless powders which increased the velocity of the bullet gradually during its travel through the barrel, brought about the additional improvements known in the revolver of today. Progressively increasing pressures permitted the refinement of the barrel rifling by shortening the pitch, thus imparting a greater spin to the bullet, resulting in greater accuracy without the fear of tumbling; while greater chamber pressures gave higher velocities and consequently flatter trajectories. Another advantage was the possibility of using harder alloys in lead bullets, and cupronickel casings for metal jacketed ones, which effectively grip the rifling without danger of stripping or excessive lead fouling.

BACON Navy Model Revolver.
Illustrated—Fig. 1, Plate 25.

Caliber .38, rim fire, six shot, single action. Sevenandonehalf inch octagonal barrel rifled with five grooves. Total length twelveandthreequarters inches. Weight 2 pounds, 2 ounces. The cylinder is oneandonequarter inches long. A steel, blade front sight is mounted on the barrel. Engraved iron frame. Spur trigger. Blued barrel and frame; — casehardened hammer. Walnut grips shellacked to a high finish.

The cylinder stops are at the front of the cylinder.

Partial removal of the cylinder permits the operation of the ejector rod, which also serves as the cylinder pin. The pin is hinged, and breaks at right angles, to facilitate use as an ejector.

The barrel of the specimen illustrated is marked "BACON MFG CO NORWICH CONN". Number 115 is stamped on the barrel, cylinder pin, and the base of the cylinder, one figure between each chamber.

The revolver was the invention of H. A. Briggs and Samuel S. Hopkins, of Norwich, Conn., Patent No. 41,117, Jan. 5, 1864. The Bacon Manufacturing Company was established in 1858 at Norwich, Connecticut, by Thomas Bacon, and was active in production of percussion pepperboxes and Briggs & Hopkins type revolvers. The firm ceased operations about 1889.

COLT ARMY REVOLVER, MODEL 1872.
Illustrated—Fig. 2, Plate 25.

Caliber .45, six shot, single action. Seven-and-one-half inch round barrel rifled with six grooves. Total length thirteen inches, weight 2 pounds, 7 ounces. The cylinder is one-and-nine-sixteenths inches long. Large, steel, blade front sight; the V-notch rear sight is in the frame. Steel frame and trigger guard. Blued barrel and cylinder, case-hardened frame and hammer. Walnut, oil finished grips.

The revolver is equipped with a rod-type hand ejector along the lower right side of the barrel. The cylinder pin is held by a set screw at the forward end of the frame. The hammer has safety, half and full cock notches. The arm was loaded and unloaded through a conventional loading gate.

The specimen illustrated is marked "COLT'S PAT. F. A. Mfg. Co. HARTFORD CT. U.S.A." The frame is marked "US" and "PAT. SEPT. 19, 1871, July 2, '72, Jan. 19, '75". The grips are marked with inspector's initials "SEB" and "PAC" on the right and left side respectively. The trigger guard and butt frame are num-bered 120775. The cylinder is marked 0775.

The revolvers of this model were submitted for test to the Ordnance Department in 1872, were adopted in 1873, 8,000 being ordered by the War Department, July 23rd that year. Additional purchases were made: 2,000 on April 30, 1874, 1,000 June 22, 1874, 1,500 July 27, 210 Dec. 2, 1874, and about 1,000 per year thereafter until 1891. The same model, but with a five-and-one-half inch barrel, was adopted for artillery use.

The original design was for a 40 grain powder, 255 grain bullet, cartridge. However since a .45 caliber Smith & Wesson using a 28 grain powder charge and a 230 grain bullet was already in the service, the same cartridge was used in the '72 Colt, and the sights of this revolver were changed accordingly, at the Frankford Arsenal, to fit either the 1872 Colt, or the Smith & Wesson, Schofield Model.

This arm was popular in the service under the name of "Peacemaker", and in the West, in forty-four caliber it became famous as the "Frontier" revolver. This heavy frame and solid construction weapon was not designed for hair-line accuracy; — there was enough clearance in the construction so that the arm operated under the most adverse conditions of field service. The large hammer spur was advantageous in the rapid draw and fire of a single action revolver, and rugged construction and solid frame made it a handy weapon in a frontier brawl when shooting was not desirable.

COLT DOUBLE ACTION REVOLVER, MODEL 1878.
Illustrated—Figs. 3 & 4, Plate 25.

Caliber .45, six shot. Seven-and-one-half inch round barrel rifled with six grooves. Total length twelve-and-one-half inches. Weight 2 pounds, 7 ounces. The cylinder is one-and-nine-sixteenths inches long. Large, steel, blade front sight; the V-notch rear sight is in the frame. Bird's head grip with checked, hard rubber grip plates. Steel trigger guard and frame. The bottom of the grip-frame carries a lanyard swivel.

The revolver is equipped with a side ejector along the lower right side of the barrel.

The specimen illustrated is marked on the top of the barrel "COLTS PT. F. A. MFG. CO. HARTFORD CT. U.S.A.". The bird-head grips are ornamented with the Colt trademark, a rampant colt in an oval medallion. Number 20,729 is stamped at the bottom of the grip frame forward of the lanyard swivel.

The double action was introduced by Colt in 1877, and made in .45 caliber in 1878. These revolvers were also made with an extra large trigger guard for Alaska service, to permit firing without removing the gauntlet. Also made with five-and-one-half inch barrel. Many of these revolvers were reissued to the service during the Philippine Insurrection, to replace the later model .38 caliber revolvers, which proved inadequate for Islands service. (See Colt Models 1901-03.)

COLT ARMY REVOLVERS, MODELS 1892-94-96.
Illustrated—Fig. 5, Plate 25.

Caliber .38, six shot, double action. Six inch round barrel rifled with six grooves. Total length eleven-and-one-half inches. Weight 2 pounds, 1 ounce. The cylinder is one-and-one-half inches long. Blade steel front sight; the rear sight is cut into the frame. Steel oval trigger guard and steel grip-frame. Blued finish.

This model is the first of the solid frame, side-swing cylinder Colts to be adopted by the army. The cylinder swings to the left to load, by drawing to the rear a catch on the left side of the frame, over the trigger guard. Simultaneous extraction of all shells is accomplished manually by an ejector rod sliding through the cylinder pin, and operating a rosette shaped extractor plate seated in the cylinder base. The system was developed in 1887, and is based chiefly on improvements invented by Carl T. Ehbets of Hartford, Conn., patents, No. 303,135, Aug. 5, 1884, No. 303,827, Aug. 19, 1884, and No. 392,503, Nov. 6, 1888, and improvements of Horace Lord, also of Hartford, Patent No. 303,172, Aug. 5, 1884, all assigned to Colts.

In 1894 the Model 1892 was improved by the addition of a locking lever, which prevented the cocking of the hammer until the cylinder was aligned and locked. Revolvers of Model 1892 were altered to incorporate the improvement. A change from black to smokeless powder was made in the cartridges.

Model 1896 is identical with 1894, and the markings described below and the revolver illustrated are typical of these 1892-94-96 models.

The barrel is marked on the left side "COLT D. A. 38", and on top "COLT'S PT. F. A. MFG. CO. HART-FORD CT. U.S.A. PATENTED AUG. 5, 1884, NOV. 6, '88, MAR. 5, '95". The frame is stamped "RCA" on the left side. The cylinder catch is numbered "K-372". The grip-frame is marked on the bottom "U.S. ARMY MODEL 1896 No. 126,372", in seven narrow lines. The plain, oil finished, walnut grips are marked with inspector's initials "RAC" and "JTT", in script, in medallions on the right and left side respectively, and date "1899" is stamped on the left grip.

COLT ARMY REVOLVERS, MODELS 1901-03.

Model 1901 was identical with the Model 1896 described above, except for an addition of an oblong lan-yard swivel at the butt.

Model 1903 differs from 1901 only in a slightly reduced bore to insure better accuracy, and in a smaller and better shaped grip.

These revolvers were used in the Spanish-American War, and later in the Philippine Insurrection. In fighting the insurrectoes, and especially in the pacification of the Southern Islands of the Philippine Archipelago, inhabited by the Mohammedan Moros, these revolvers were found wanting, due to the insufficient stopping power of the 150 grain bullet against an individual, crazed with drugs and religious fanaticism. In the Islands the light .38s were replaced with the old .45s, whose heavy slug stopped and dropped the juramentado, bound by an oath to win his way to the Mohammedan heaven by chopping down as many non-believers as possible, before being killed himself. These fanatics underwent a special ceremony of purification before running amuck, and were known to have kept swinging a kris or campilan, with six or more bullets in the body. Tradition has it that General Pershing, then captain and one of the Southern Islands administrators, succeeded in discouraging the self-dedication for heavenly attendance, by burying killed juramentadoes with a butchered pig, to a Mohammedan, an unclean animal. Association with swine effectually denied the luscious, or shall we say lascivious, rewards of the Islamic Prophet's heaven earned for mass murder of Christians, and eventually succeeded in stamping out the practice of juramentado.

COLT NEW NAVY REVOLVERS, MODELS 1889-1895.

A caliber .38, side swing, double action revolver with a six inch barrel rifled with five grooves, and similar in appearance and action to the Army 1892-94-96 Models described above, was brought out by Colts in 1889. This arm was improved in 1895 by a locking lever device which locked the hammer and trigger when the cylinder was not in line with the barrel.

The revolvers were stamped on the barrel with the usual Colt marking and patents dates, and were marked on the butt frame "U.S.N. 38 D.A.", the number of the arm and the year of the model.

COLT MARINE CORPS, MODEL 1907.

Caliber .38, six shot, double action. Six inch round barrel rifled with five grooves. Total length ten-and-three-quarters inches. Weight 2 pounds, $\frac{1}{2}$ ounce. Butt equipped with lanyard ring. Checked walnut grips.

Except for the smaller grip, rounded at the rear, this arm is similar in appearance and action to the Navy Model 1895 and the Army Model 1903. Typical marking: barrel—"COLT'S PT F. A. MF'G CO. HARTFORD CT. U.S.A. PATENTED AUG. 5, 1884, NOV. 6, '88, MAR. 5, '95"; butt frame—"U.S.M.C." and number.

COLT NEW SERVICE REVOLVER, MODEL 1909.
Illustrated—Fig. 6, Plate 25.

Caliber .45, six shot, double action. Five-and-one-half inch round barrel rifled with five grooves. Total length ten-and-five-eighths inches. Weight 2 pounds, 8 ounces. The fluted cylinder is one-and-five-eighths inches long. Steel blade front sight; rear sight in the frame. Blued barrel, cylinder and frame; bright hammer; case-hardened trigger. Oil finished walnut grips. Lanyard swivel in butt.

Side swing action operated by a catch on the left side of the frame, behind the cylinder. Simultaneous extraction of all shells is accomplished by an ejector-rod sliding through the cylinder and operating a rosette-shaped extractor plate, seated in the cylinder base.

The specimen illustrated is marked on the top of the barrel "COLT'S PT. F. A. MFG. CO., HARTFORD CT. U.S.A. PT. AUG. 5, 1884, JUNE 5, 1900, JULY 4, 1905", and on the side "COLT D.A. 45". The frame is stamped with the Colt trademark of a rampant colt. The right of the frame and the grips are stamped at the base with inspector's initials "RAG". The butt frame is marked "U.S. ARMY MODEL 1909 No. 30294".

This revolver modified by a smaller grip with lanyard ring and checked walnut stock, was adopted by the U. S. Marine Corps. Marked "U.S.M.C." on the butt frame.

COLT ARMY REVOLVER, MODEL 1917.
Illustrated—Fig. 1, Plate 26.

Caliber .45, six shot, double action. Five-and-one-half inch round barrel rifled with five grooves. Total length ten-and-three-quarters inches. Weight 2 pounds, 8 ounces. The fluted cylinder is one-and-seventeen-thirty-seconds inches long. Steel blade front sight; rear sight in frame. Dull blued barrel frame and cylinder, bright hammer. Oil finished walnut grips.

This side-swing revolver, like the Model 1917 Smith & Wesson, used .45 caliber rimless automatic pistol ammunition Model 1911, fitted three cartridges to a thin steel clip. These revolvers were made to supply a war need due to shortage of automatic pistols, and to fill an acute demand for hand arms. They were designed to take the rimless automatic pistol cartridge to avoid an additional type of ammunition to be supplied to troops in field. The problem of ejection of the rimless shells was met by the provision of the slotted semi-circular steel clips, which held the shells at the groove of the cartridge base, and permitted the action of the ejector against the clip, extracting the six shells and the two clips simultaneously. The clips could be used over again. The revolver can also be used with automatic type rimmed cartridges.

The specimen illustrated is marked on the top of the barrel "COLT'S PT. F. A. MFG. CO., HARTFORD CT., U.S.A., PATENTED AUG. 5, 1884, JUNE 5, 1900, JULY 4, 1905," on the side "COLT D. A. 45" and under the barrel "UNITED STATES PROPERTY". The frame is stamped with the Colt trademark. The butt is marked "U.S. ARMY MODEL 1917 No. 95268".

151,700 Colt Model 1917 Revolvers were purchased by the U. S. Government between April 6, 1917 and December, 1918, on World War contracts.

FOREHAND & WADSWORTH Army Revolver.
Illustrated—Fig. 2, Plate 26.

Caliber .44, six shot, single action. Seven-and-one-half inch round barrel, rifled with six grooves. Total length thirteen-and-one-eighth inches. Weight 2 pounds, 8 ounces. Large, steel blade front sight; the rear sight is in the frame.

The ejector pin is attached to the barrel by a thin band which encircles the barrel. Forward movement on a knurled stud in front of the frame, permits the ejector to move forward, and a rotation of one-quarter of a turn allows it to clear the frame and enter the cylinder chambers. Rotation of the ejector also gives access to the cylinder pin. The cylinder is loaded and unloaded by a conventional loading gate.

The specimen illustrated is marked on the barrel "FOREHAND & WADSWORTH, WORCESTER, MASS. U. S. PAT'D OCT 22 '61 JUNE 27 '71 OCT 28 '73". Number 74 is marked on the butt of the grip frame, and on the cylinder pin journal.

The revolver was made about 1877 by S. Forehand and H. C. Wadsworth, sons-in-law of Ethan Allen, of the firm of Allen & Wheelock, arms manufacturers. Upon Mr. Wheelock's death in 1863, Forehand and Wadsworth were admitted to partnership in the firm, and in 1875 the firm of Allen & Wheelock became Forehand &

Wadsworth. In 1902 the firm sold out to Hopkins & Allen of Norwich, Conn., which during the late war was absorbed by the Marlin-Rockwell Corporation.

FOREHAND & WADSWORTH NEW ARMY REVOLVER, (No. 44.)

Caliber .44, six shot, single action. Six-and-one-half inch round barrel, rifled with six grooves. Weight 2 pounds, 8 ounces. Wrought iron frame. Blade front sight, V-notch rear sight in the frame. Fired the Winchester center fire cartridge Model 1873, of 40 grains of powder and 200 grain bullet. Lighter Smith & Wesson Russian cartridge could also be used.

The revolver is similar to the old model described above, but the weak ejector pin of the old model has been replaced by a side ejector on the lower right side of the barrel. The cylinder pin is exposed. The revolver was equipped with a safety notch, as well as the usual half cock notch. The specifications of the improved model, were advertised in a circular in March, 1878.

HOPKINS & ALLEN NAVY REVOLVER, XL MODEL.
Illustrated—Fig 3, Plate 26.

Caliber .38, rim fire, six shot, single action. Six inch round barrel, rifled with five grooves. Total length eleven inches. Weight 1 pound, 10 ounces. Fluted cylinder. Brass blade front sight; rear sight in frame. This specimen is nickel-finished except the hammer and trigger, which were case-hardened.

Pressure on a catch in front of the frame permits the sideward movement of the ejector.

The revolver is marked on top of the barrel "HOP-

KINS & ALLEN MFG. CO., PAT. MAR. 28, '71, APR. 27, '75" and on the barrel strap "XL NAVY". The barrel is stamped underneath "38-100 CA."

MERWIN & HULBERT Army Revolver, MODEL 1876.
Illustrated—Fig. 4, Plate 26.

Caliber .44, six shot, single action. Seven inch round barrel, rifled with five grooves. Total length twelve inches. Weight 2 pounds, 9-1/2 ounces. The fluted cylinder is one-and-nine-sixteenths inches long. Large blade front sight; rear sight in frame. Steel frame and trigger guard. Bird's-head crested handle with checked, hard rubber grips and a lanyard hole through the crest of the butt frame. Nickel-plated finish, except the hammer and trigger, which are case-hardened.

The revolver has no top-strap. The loading gate slides down and permits loading from the right side of the frame. To open the revolver, the arm is half-cocked, the milled projection under the frame is moved towards the trigger guard, and a catch on the left side of the barrel lug pushed in. The barrel may then be turned to the left and drawn forward with the cylinder, permitting easy extraction.

The specimen illustrated is marked on the top of the barrel "MERWIN HULBERT & CO., NEW YORK, U.S.A., PAT. JAN. 24, APR. 21, DEC. 15, '74, AUG. 3, '75, JULY 11, '76, APR. 17, '77, PAT'S. MAR. 6, '77". On the left side of the barrel is stamped "HOP-KINS & ALLEN MANUFACTURING CO., NOR-WICH, CONN., U.S.A." The frame is marked on the left "CALIBRE WINCHESTER 1873".

This revolver was also made in double action and some were made with conventional square butt. The arm was tested by the government in 1876, and rejected.

MERWIN & HULBERT ARMY POCKET REVOLVER.
Illustrated—Fig. 5, Plate 26.

Caliber .44, six shot, double action. Three-and-five-sixteenths inch round barrel, rifled with five grooves. Total length eight-and-five-eighths inches. Weight 2 pounds, 5 ounces. Large blade front sight; rear sight in frame. Steel frame and trigger guard. Bird's-head shaped, crested handle with checked, hard rubber grips, and a lanyard hole through the crest of the butt frame.

The loading gate slides down and permits loading from the right side of the frame. To open the revolver, the arm is half-cocked, the milled projection under the frame is pulled towards the trigger guard, and a milled catch, on the left side of the barrel lug is pushed in. The barrel may then be turned to the left and drawn forward with the cylinder, permitting easy extraction. All the metal parts of the specimen illustrated and described were nickel-plated. The revolver differs from the Army Model 1876 chiefly in the top strap and shorter barrel.

The revolver is marked on the right side of the frame "MERWIN HULBERT & CO., N. Y. POCKET ARMY", and on the left side "CALIBRE WINCHES-TER 1873". The left side of the barrel is marked "HOP-KINS & ALLEN M'F'G CO., NORWICH, CONN., U.S.A. PAT. JAN. 24, APRIL 21, DEC. 15, '74, AUG. 3, '75, JULY 11, '76, APR. 17, '77. PATS. MAR. 6, '77."

PLANT ARMY REVOLVER, PERCUSSION OF CUP-PRIMER CARTRIDGE.

Illustrated—Fig. 1, Plate 27.

Caliber .42, six shot, single action. Six inch octagonal ribbed barrel rifled with six grooves. Total length ten-and-three-quarters inches. Weight 2 pounds. White brass front sight set in the rib, rear sight in frame. The frame is bronze, silver plated. Sheath trigger. Rosewood grips.

A spring-held cylinder pin is withdrawn from the front to remove the cylinder for loading. The revolver used a hollow base cup-primer cartridge loaded from the front of the cylinder, (to avoid infringement of the Smith & Wesson and Rolin White patents). The hammer tip strikes the inside concave portion of the copper base of the cartridge which contains the fulminate. The empty shells are ejected by a side ejector on the right side of the frame, below the hammer.

This revolver could also be fired as a percussion arm. Extra interchangeable cylinders were obtainable for use when cup primer cartridges were not available. These percussion cylinders have recessed primer cones.

The specimen illustrated is marked "PLANT'S MFG. CO., NEW HAVEN CT". on the barrel-rib, and "MER-WIN & BRAY, NEW YORK", on the left side of the barrel. The cylinders are stamped "PATENTED JULY 12, 1859 & JULY 21, 1863".

The revolver is the invention of Willard E. Ellis and N. White, and was patented by them July 12, 1859, July 21, 1863 and Aug. 25, 1863, Patent Numbers 24,726,— 39,318 and re-issue 1,528. The patents were assigned to Ebenezer H. Plant, Henry Reynolds, Amzi H. Plant and Alfred Hotchkiss.

The revolver was distributed by the Merwin & Bray Company to supercede the Prescott revolver, which was an infringement of the Smith & Wesson patents. Merwin & Bray are not known to have manufactured arms, but did finance a number of arms companies, whose products carried the Merwin & Bray name.

The makers of the revolver, Plant Mfg. Co., who made carriage hardware, were burned down in 1866, and it is reported that J. J. Marlin continued the manufacture. The name of the arm was later changed to Eagle revolver, a product of the Eagle Arms Company, which was incorporated in New York, Nov. 20, 1865.

POND ARMY REVOLVER.
Illustrated—Fig. 2, Plate 27.

Caliber .44, rim fire, six shot, single action. Seven-and-one-quarter inch octagonal barrel, rifled with five grooves. Total length twelve-and-three quarters inches. Weight 2 pounds, 8 ounces. The cylinder is one-and-seven-sixteenths inches long. Steel blade front sight; V-notch rear sight in the frame. Sheath trigger, shellacked walnut grips. This specimen is silver-plated, except the hammer and trigger which were case-hardened.

Pressing two lugs at the lower part of the frame, tips the barrel upwards on a hinge in front of the hammer.

The barrel of the specimen illustrated is marked "L. W. POND, WORCESTER, MASS."

The revolver, manufactured by Lucius W. Pond under Patent No. 35,625 of June 17, 1862, was an infringement on the Smith & Wesson patents, and 4,486 Pond revolvers were turned over to Smith & Wesson in 1863, in settlement of the infringement.

PRESCOTT NAVY REVOLVER.
Illustrated—Fig. 3, Plate 27.

Caliber .38, rim fire, six shot, single action. Seven-and-five-sixteenths inch octagonal barrel, rifled with five grooves. Total length twelve-and-one-half inches. Weight 1 pound, 13 ounces. The cylinder is one-and-one-quarter inches long. Brass blade front sight; the V-notch rear sight is in the frame, which is made of brass. Blued barrel and cylinder. The almost circular trigger guard forks at the rear. Polished rosewood grips.

The cylinder may be removed for loading, by pressing a stud on the cylinder-pin, and withdrawing the pin forward. The extraction of the .38 long, rim fire shells was performed manually, with a nail or pencil.

The specimen illustrated is marked on the barrel "E. A. PRESCOTT, WORCESTER, MASS., PAT. OCT. 2, 1860".

The revolver was invented by E. A. Prescott, Patent No. 30,245, Oct. 2, 1860, and manufactured at Worcester, Mass. It was advertised in 1862 by the firm's sales agents, Merwin & Bray of 245 Broadway, New York City, as "8 inch Navy size, carrying a ball weighing 38 to the pound." Prescott was sued by Smith & Wesson for infringement of their Rolin White patents (See Smith & Wesson).

The arm was made in a number of barrel lengths, finishes and variations. Six-and-one-eighth and seven-and-one-eighth inch barrels are known. Some revolvers were made with steel frames, others had nickeled finish, and some were loaded through a slot in the recoil shield.

REMINGTON Army Revolver, NEW MODEL 1874.
Illustrated—Fig. 4, Plate 27.

Caliber .44, six shot, single action. Seven-and-one-quarter inch round barrel, rifled with five grooves. Total length twelve-and-seven-eighths inches. Weight 2 pounds, 11 ounces. The fluted cylinder is one-and-seventeen-thirty-seconds inches long. A German silver, blade front sight is mounted on the barrel; the rear sight is in the frame. Blued barrel, cylinder, frame, trigger and steel trigger guard. The hammer and the loading gate are case-hardened in mottled colors. Oil finished walnut stocks. Lanyard ring in butt.

The revolver is similar in appearance to the Colt "Peacemaker" Model 1872. The arm is equipped with a side rod hand ejector, along the lower right of the barrel, and a conventional loading gate on the right side. It was first designed for a 28 grain powder, 227 grain ball cartridge, which was later superceded by the Winchester 44-40.

The barrel of the specimen illustrated is marked "E. REMINGTON & SONS, ILION, N.Y., U.S.A." The left grip is stamped with inspector's initials "FR".

3,000 of these revolvers were issued to the army in 1875, and 3,000 more, slightly modified, later.

SMITH & WESSON Army Revolver No. 2.
Illustrated—Fig. 5, Plate 27.

Caliber .32, rim fire, six shot, single action, using .32 caliber long S. & W. cartridge. Six inch octagonal, ribbed barrel rifled with five grooves. Total length ten-and-three quarters inches. Weight 1 pound, 8 ounces. The cylinder is one-and-three-sixteenths inches long. A white brass front sight is set into the barrel rib; the rear sight, a small V-shaped groove is cut into the top of the cylinder stop. Sheath trigger. Full cock notch only. Blued barrel, frame and cylinder. Rosewood grips. Also made in nickel finish.

The revolver is of the top-break system, hinged in rear of the barrel, over the front of the cylinder. The cylinder stop is in the frame extension strap over the cylinder. Removal of the cylinder pin permits the use of the pin for the extraction of shells.

The revolver illustrated is marked on the barrel "SMITH & WESSON, SPRINGFIELD, MASS." The cylinder is marked "PATENTED APRIL 3, 1855, JULY 5, 1859 and DEC. 18, 1860". No. 30,121 is stamped on the butt.

76,502 of these revolvers were manufactured from June, 1861 to 1874. Though not adopted by the government, these arms were popular as personal or pocket side-arms with the officers of the Union forces during the Civil War.

The arms manufacturing firm of Smith & Wesson of Springfield, Mass., originated in the partnership of Horace Smith, maker of rifle barrels, and Daniel B. Wesson, younger brother of Edwin Wesson, co-inventor of the WESSON & LEAVITT revolver. Their association began while both were employed by Allen, Brown and Luther, makers of musket and rifle barrels. (Frederic Allen, Andrew J. Brown and John Luther of Worcester, Mass.)

At about 1854 Smith & Wesson engaged at Nor-wich, Conn., in the manufacture of a magazine firearm, the patent for which had been issued to Smith in 1851. Wesson's possible contribution to the firm, patents to a mechanically operating revolver, which he inherited in part from his brother Edwin, had to be held in abeyance as the result of a successful suit brought by Colts against Massachusetts Arms Company, formed by the Wesson heirs.

About 1855, Smith & Wesson sold out their patent rights and machinery to Oliver D. Winchester, who moved the plant to New Haven and organized the Vul-canic Arms Company. After the sale of their magazine arm, Smith & Wesson moved to Springfield, Mass., and prepared to manufacture revolvers under the Horace Smith, Horace Smith & Daniel B. Wesson, and the Rollin White patents, upon the expiration of the Colt patents in the fall of 1856.

The first Smith & Wesson revolver produced by the firm was a small caliber .22 arm, using a metallic, copper shell cartridge. The small caliber was due to the difficulty of production of a copper case, strong enough not to rup-ture when fired, yet thin enough to be dented by the hammer and set off the detonating compound in the base of the shell. With the development of better methods of production and annealing of copper shells, resulting in greater tensile strengths necessary because of the lack of outside support in the early revolver cartridges, Smith & Wesson brought out a .32 caliber rim fire cartridge revolver, many of which were carried as personal or pocket weapons by the officers of the Union Army during the Civil War.

In the meantime, Smith & Wesson, through their cartridge revolver patents acquired by the purchase of the Rollin White's invention of a "cylinder bored end to end", held the virtual monopoly of cartridge revolver manufacture until 1869, when the White patents expired. By promise of infringement suits in some cases, and by court action in others, Smith & Wesson had obtained the cessation of manufacture and judgments, against their competitor manufacturers of cartridge revolvers, who infringed on their patents.

It was not until 1869, however, that Smith & Wesson were able to manufacture a large caliber revolver, as the weak latch of their top-break construction would not stand the higher pressures of the heavier loads. In 1869, Smith & Wesson purchased the W. C. Dodge patents Nos. 45,912 and 45,983 for a stronger latch and improved hinge, and the Charles A. King patent No. 94,003 for simultaneous extraction of all shells by a rack set flush with the rear of the cylinder.

The result of these acquisitions was their first martial, caliber .44, American Model of 1869, a top-break revolver. In 1875 this was followed by the caliber .45 Schofield Model, similar to the American in action and appearance. The next martial Smith & Wesson to be used by the armed services was the side swing Model 1899 in caliber .38, and the last, the justly famous caliber .45 Model 1917 Army, used in the World War.

SMITH & WESSON ARMY REVOLVER, MODEL 1869, AMERICAN.

Illustrated—Fig. 1, Plate 28.

Caliber .44, six shot, single action. Eight inch round,

ribbed barrel, rifled with five grooves. Total length thir-teen-and-one-half inches. Weight 2 pounds, 10-1/2 ounces. The fluted cylinder is one-and-seven-sixteenths inches long. German silver, blade front sight; the rear sight is on the latch. Blued barrel, frame, cylinder and trigger; case-hardened hammer. Oil finished walnut stocks.

The revolver is on the top-break system, operated by a latch in front of the hammer. The barrel is jointed to the frame at the forward end of the bottom strap. The ejector is operated by a rack and pinion mechanism, incor-porated in the barrel joint. Breaking the revolver causes the ejector, which is set on a square shaft, to rise in rear of the cylinder and throw out all the empty shells in one operation. The arm used S. & W. American, center fire cartridges with a 218 grain, outside lubricated, round nose bullet and a 28 grain black powder charge.

The revolver illustrated is marked on the barrel "SMITH & WESSON, SPRINGFIELD, MASS., U.S.A., PAT. JULY 10, 1860, JAN. 17, FEB. 17, JULY 11, '65, AUG. 24, 1869". Number 16,802 is marked on the frame.

This first large caliber model to be manufactured by Smith & Wesson, was the result of improvements based on the W. C. Dodge patents of Jan. 17 and 24, 1865, Patents Nos. 45,912 and 45,983 for a stronger latch and improved hinge, and the Charles A. King Patent No. 94,003 of Aug. 24, 1869, for simultaneous ejection by rack and pinion.

1,000 of these revolvers were ordered by the army Dec. 29, 1870, and were delivered in 1871.

SMITH & WESSON ARMY REVOLVER, MODEL 1875, SCHOFIELD.

Illustrated—Figs. 2 & 3, Plate 28.

Caliber .45, six shot, single action. Seven inch round, ribbed barrel rifled with five grooves. Total length twelve-and-one-half inches. Weight 2 pounds, 8 ounces. The fluted cylinder is one-and-seven-sixteenths inches long. A steel, flat top, blade front sight is set into the rib. The rear sight is in the latch. Blued barrel, frame, cylinder and trigger. Case-hardened hammer. Oil finished walnut stocks.

The revolver is similar in appearance to the Model 1869, American S. & W. The top-break system operates by a latch on the receiver, the latch also forming the rear sight. The ejector is set on a rounded shaft and rises in rear of the cylinder, when the revolver is broken.

Later models were improved by a milled top-latch with a hollow rear, to facilitate the placing of the thumb for cocking. The revolvers used a .45 caliber S. & W. cartridge with 230 grain bullet and 28 grains of black powder.

The barrel of the specimen illustrated is marked on the left side "SMITH & WESSON, SPRINGFIELD, MASS., U.S.A., PAT. JAN. 17th & 24th, '65, JULY 11th, '65, AUG. 24th, '69, JULY 25th, '71", and on the right side "SCHOFIELD'S PAT. APR. 22d, '73". Number 1532 and "US" are stamped on the butt frame.

The revolver embodied the George W. Schofield, U.S. Army, patents No. 116,225, June 20, 1871, and No. 138,041, April 22, 1873, for auto ejection.

The arm was submitted to the Ordnance Department for tests in 1872, was accepted, and 6,000 were made from 1875 and supplied to the army; 3,000 under contract of March 15, 1876, and 3,000 in 1877. It is believed that all of the latter issue incorporated the improved hollowed latch.

SMITH & WESSON Navy Revolver, MODEL 1881.
(New Model Navy No. 3.)

Illustrated—Fig. 4, Plate 28.

Caliber .44, six shot, double action. Six-and-one-half inch round, ribbed barrel rifled with five grooves. Total length eleven-and-one-half inches. Weight 2 pounds, 5 ounces. The fluted cylinder is one-and-nine-sixteenths inches long. A steel blade front sight is set into the rib. The rear sight is in the latch. Blued barrel, frame and cylinder; case-hardened hammer and trigger. Checked walnut grips. The rear of the trigger guard is squared.

The revolver is similar in appearance to the American and Schofield models. The top break system operates by a latch in the receiver. The ejector is set on a hexagonal shaft and rises in rear of the cylinder when the revolver is broken. The revolver used the .44 caliber S. & W. Russian cartridge.

The specimen illustrated is marked on the top of the barrel rib "SMITH & WESSON, SPRINGFIELD, MASS, U.S.A., PAT. JAN. 17th & 24th, '65, JULY 11th, '65, AUG. 24th, '69, JULY 25th, '71, DEC. 2nd, '79, MAY 11th, 1880, MAY 25th, 1880"; and on the right side with the S. & W. monogram trademark.

Though called a Navy revolver in the early S. & W. catalog the arms were not officially adopted by the service. They were also made in four, five and six inch bar-rel lengths, and in long and short strap models, the latter using a one-and-seven-sixteenths inch cylinder. The finish was blue or nickel, made in both checked walnut and hard rubber grips, with the S. & W. monogram in grip circles.

Numbers 1 to 54,668 were manufactured from 1881 until 1913, of this number 275 were made with 6-1/2 inch barrels and 1-9/16 inch cylinders for the .38-40 caliber Winchester rifle cartridge.

A similar arm was also brought out called "New Model Army No. 3", which likewise was not adopted by the military service. The revolver was similar to the Navy Model described above, except that the arm was single action and the trigger guard was oval. It was made in six and in six-and-one-half inch barrel lengths only.

SMITH & WESSON ARMY, NAVY REVOLVER, MODEL 1899. Illustrated—Fig. 5, Plate 28.

Caliber .38, six shot, double action. Six inch round barrel rifled with five grooves. Total length eleven inches. Weight 1 pound, 15 ounces. The fluted cylinder is one-and-nine-sixteenths inches long. Steel blade front sight; rear sight in the frame. Blued barrel, frame and cylinder; case-hardened hammer and trigger. Checked walnut grips.

The left swing cylinder is operated by a roughened catch set in the left side of the frame, back of the cyl-inder. Hand ejector. Rebounding hammer. The arm used the .38 caliber Colt U. S. service cartridge.

The specimen illustrated is marked on the top of the barrel "SMITH & WESSON, SPRINGFIELD, MASS.,

U.S.A., PAT'D JULY 1, '84, APRIL 9, '89, MAY 21, '95, AUG. 4, '96, DEC. 22, '96, OCT. 4, '98", and on the side "S. & W. 38 MIL." The frame is marked on the left side "K. & M", and on the right side with the Smith & Wesson trade mark, initials "SW" in a medallion. The grip-frame is marked "U. S. ARMY MODEL 1899". Number 13845 is stamped on the forward side of the grip-frame. The checked grips are stamped with inspector's initials "JTT" and "KSM", left and right respectively, and the left grip is dated 1901.

2,000 were supplied to the navy and 1,000 to the army. After 1899 the revolver was made for many years in a number of minor variations.

The navy revolvers of this model were marked on the butt "S. & W. NAVY 1899 U.S.N.", an anchor, "38 D.A." an arrow, and the number of the revolver, as well as the initials of the inspector, such as "JAB".

SMITH & WESSON ARMY REVOLVER, MODEL 1917.
Illustrated—Fig. 6, Plate 28.

Caliber .45, six shot, double action. Five-and-seven-sixteenths inch round barrel rifled with six grooves. Total length ten-and-three-quarters inches. Weight 2 pounds, 4 ounces. The fluted cylinder is one-and-seventeen-thirty-seconds inches long. Steel blade front sight; U-notch rear sight in frame. Smoothly blued barrel, cylinder and frame. Case-hardened hammer and trigger. Oil finished walnut grips. Lanyard loop at base of the butt.

This side-swing revolver like the Model 1917 Colt, used caliber .45 rimless automatic cartridges, Model 1911, loaded in two clips, three rounds to a clip. The revolvers were made to supply a war need due to shortage of automatic pistols, and were designed to take the rimless automatic pistol cartridges to avoid an additional type of ammunition to be supplied to the troops. The problem of ejection of the rimless cartridge was solved by the slotted, semi-circular, thin, steel clips in which the cartridges were held by the slot in the base of the groove, the clip providing the abutment, against which the ejector could act to eject the six empty shells and the two clips simultaneously. Though the revolver would fire without the clips, extraction would have to be performed manually, by a small rod or pencil, or the shells pulled out by finger nails, or the rim of another shell.

The revolver illustrated is marked on the top of the barrel "SMITH & WESSON, SPRINGFIELD, MASS., U.S.A., PATENTED DEC. 17, 1901, FEB. 6, 1906, SEPT. 14, 1909", on the side "S. & W. D. A. 45", and on the bottom "UNITED STATES PROPERTY". The butt of the frame is marked "U. S. ARMY MODEL 1917 No. 160976".

153,311 S. & W. Model 1917 revolvers were purchased by the U. S. Government between April 6, 1917, and December, 1918, for use of the armed forces during the World War.

Chapter 3.

AUTOMATIC PISTOLS.

AUTOMATIC PISTOLS.

Towards the end of the Nineteenth Century inventors began to take advantage of the recoil power of the explo-sion of the cartridge. The result of their efforts were a number of automatic and autoloading arms.

Automatic firearms are those in which the force of the recoil of the first shot is utilized to continue the operation of unlocking, extraction, ejection, feeding, loading, locking and firing continuously as long as the ammunition lasts in the magazine, belt or strip, or the sear is kept free of the operating mechanism by pressure on the trigger. Because of necessity of a fixed rest or support, automatic arms are principally machine guns and machine rifles, and the majority of these arms are gas as well as recoil operated.

In pistols the recoil of the cartridge is used to make the arm autoloading, that is, the force of the explosion of each shot is used to unlock the mechanism, extract and eject the empty shell, and reload, by stripping and feeding another cartridge from the magazine into the chamber. The trigger must be pressed for each successive shot. Automatic fire in an unsupported hand arm would result in a wild, uncontrolled burst of fire, the pistol climbing up and away after the first shot.

Autoloading pistols gained rapid and well deserved popularity at the beginning of the Nineteenth Century. As compared with the revolver, the autoloading pistol is less cumbersome, harder shooting, for there is no possibility of gas escape between the cylinder and the barrel, the arm can be fired more rapidly and more accurately than a double action revolver, and has the advantage of greater magazine capacity.

Though the name "automatic" as applied to autoloading pistols is a misnomer, these pistols are so called in this chapter because of the accepted use of the name by the public and by the War Department.

COLT Automatic Pistol, MILITARY MODEL 1902.
Illustrated—Fig. 1, Plate 29.

Caliber .38, recoil operated, magazine fed, autoloading. The pistol has capacity for eight shots, rimless automatic type cartridges. The barrel is six inches long and is rifled with six grooves. The slide is eight-and-three-eighths inches long. Total length of the pistol is eight-and-seven-eighths inches. Weight 2 pounds, 5 ounces. A steel blade front sight and a U-notch rear sight are mounted on the slide, which is grooved at the forward end. The hammer is of the striker, or stub type, without a spur. There are no safety devices. The hard rubber, checked grips are marked "COLT", and with the Colt trademark, a rampant colt in a circle. Blued finish.

The functioning of the pistol is similar to that of the Model 1911, fully described in the following pages. In brief, when the cartridge is fired, the barrel and slide move rearward for a short distance, when the barrel stops and drops, unlocking the slide, which continues to the rear, extracting and ejecting the empty shell. When the limit of rearward travel has been reached, the slide returns, strips a new cartridge from the magazine and drives it into the chamber. The forward movement of the slide forces the barrel forward and upward, locking it to the slide, ready for firing upon pressure on the trigger.

The pistol is marked on the right side of the slide "AUTOMATIC COLT CALIBRE 38 RIMLESS SMOKELESS". On the left side of the slide is marked "BROWNING'S PATENT PAT'D APRIL 20, 1897, SEPT. 9, 1902 COLT'S PATENT FIRE ARMS MFG. CO., HARTFORD, CONN., U.S.A." and is stamped with a rampant colt in a circle. Number 6,173 is stamped on the left side of the frame above the trigger guard.

The pistol is the invention of John M. Browning of Ogden, Utah, a famous firearms inventor.

COLT AUTOMATIC PISTOL, MODEL 1911.
Illustrated—Fig. 2, Plate 29.

Caliber .45, recoil operated, magazine fed, self loading. The magazine has capacity for seven rounds of rimless, alloy jacketed ammunition, Model 1911. The barrel is five inches long, and is rifled with six grooves making one left turn in sixteen inches. The total length of the pistol is about eight-and-one-half inches. Weight of the pistol with empty magazine is 2 pounds, 7 ounces. A blade front sight and a U-notch rear sight are mounted on the slide. Checked walnut stocks. Blued finish.

The pistol functions as follows: When the first cartridge is fired, the force of the recoil is utilized to drive the slide and the barrel to the rear. The action of the barrel link stops the barrel and swings it downward, unlocking it from the slide. The slide continues to move to the rear, extracting and ejecting the empty cartridge case, and simultaneously compressing the recoil spring, until the slide reaches its rearmost position. It is then driven forward by the action of the compressed recoil spring. In its forward movement the slide strips from the magazine a cartridge which has been raised to protrude slightly in front of the slide by the action of the magazine spring. The slide then forces the cartridge into the chamber, and at the same time drives the barrel, which is pivoted on the link, forward and upward, locking the barrel to the slide by means of two transverse ribs on the barrel fitting into corresponding slots on the slide. The pistol is then ready to fire again, upon pressing the trigger.

The pistol fires but once on each squeeze of the trigger. When the last cartridge has been fired, the slide remains open. If the magazine catch is depressed, the empty magazine will fall out, and a loaded magazine can be inserted, making seven more shots available. The slide is returned forward, and the chamber loaded by downward pressure on the slide stop, the hammer remaining cocked and ready for firing.

The pistol is equipped with three safety devices as follows:—

(1) A safety lock on the left side of the frame, in front of the hammer. When the safety is raised while the hammer is cocked, the latter is locked in that position.

(2) A grip safety which must be compressed into the grip by the yoke of the hand while the trigger is being squeezed.

(3) A disconnector mounted inside the receiver in rear of the magazine seat. The disconnector is depressed by the slide in all positions except when the slide is fully forward and locked to the barrel. In the depressed position of the disconnector the trigger is disconnected from the sear, permitting the sear to re-engage the hammer. This arrangement automatically and positively prevents the firing of the pistol except when all its parts are closed and locked in firing position; it also prevents more than one shot being fired following each squeeze of the trigger.

About 1926 the Colt Company introduced certain improvements in the pistol, such as a longer grip safety tang to protect the hand from hammer pinching, a shorter knurled trigger-face, a raised knurled trigger-housing to fit the hand better and prevent slippage, a clearance cut made on the receiver for the trigger-finger, and lastly a widened front sight and a corresponding rear sight. Of these improvements, the first three parts are interchangeable with the old. As improved the pistol is called Model 1911 A1.

The pistol as made by Colt is marked on the left side of the slide "COLT'S PAT. F. A. MFG. CO., HARTFORD CT., U.S.A.", a rampant colt, and "PATENTED APRIL 20, 1897, SEPT. 9, 1902, DEC. 19, 1905, FEB. 14, 1911, AUG. 19, 1913", and on the left side of the frame "UNITED STATES PROPERTY". On the right side the slide is marked "MODEL OF 1911 U. S. ARMY", and on the frame the specimen described bears the number 159598.

The pistol is the invention of Mr. John M. Browning, a noted firearms inventor, famed for the machine guns and automatic rifles and shotguns made from his designs. The pistol rendered a splendid account of itself during the World War, for in the hands of pistol shooting American troops it was a deadly weapon as compared to the smaller calibers relied on by officers of foreign armies, who regarded the pistol as more of an ornamental side-arm than a military weapon.

Though at first but few men of an infantry regiment carried pistols, the effectiveness of the arm in trench and close fighting proved the desirability of more extensive issue, and created an enormous demand that exceeded the capacity of the Colt Company facilities, as well as the facilities of the Springfield Armory, where these pistols had also been manufactured, but which were now strained to meet the demand for rifles.

Through the cooperation of the Colt Company, drawings and plans were made available to the Remington-UMC Company, whose production augmented the Colt output, all parts of the pistol of both companies being interchangeable.

In 1918, in order to fill the constantly growing pistol requirements of the American Expeditionary Forces, contracts for these pistols were given to the National Cash Register Company of Dayton, Ohio; The North American Arms Co. of Quebec, and Caron Bros. of Montreal, Canada; The Savage Arms Company, Utica, N. Y.; Burroughs Adding Machine Company, Detroit, Mich.; The Winchester Repeating Arms Company of New Haven, Conn.; The Lanston Monotype Co. of Philadelphia, Pa., and the Savage Munitions Co. of San Diego, California.

The coming of the Armistice resulted in the cancella-
tion of the contracts before production began, and the
only pistols obtained during the World War were made
by Colts and Remington Arms.

At the outbreak of the war the army had approximately 75,000 automatic pistols in storage and in hands
of troops. At the signing of the Armistice this number had
grown to 643,755. Between April 6, 1917, and December 1918, the Colt Company produced 425,000 automatic pistols, and the Remington-UMC, who did not
begin production until September, 1918, made 13,152.

GRANT-HAMMOND AUTOMATIC PISTOL.
(NAVY EXPERIMENTAL.)
Illustrated—Fig. 3, Plate 29.

Caliber .45, recoil operated, magazine fed, self loading. The pistol has capacity for eight rounds of rimless
caliber .45 automatic cartridges. The barrel is six-and-
three-quarters inches long and is rifled with six grooves.
Total length eleven-and-one-quarter inches. Weight of the
pistol with empty magazine is 2 pounds, 10 ounces. Part-
ridge type band front sight; the U-notch rear sight is
mounted on the slide. Outside hammer. Safety lock on
the left side of the frame. Smooth grips. Blue finish.

The barrel is screwed into the slide, inside of which
is the breech block. Cocking handles similar to the Mauser
automatic pistol type, extend across the end of the block.
When the pistol is discharged, the recoil causes the locked
slide and block to travel together for a short distance,
when the block is unlocked from the slide, and continues
to the rear, extracting and ejecting the empty shell. On

return to forward position the block strips a live cartridge from the magazine feedway and drives it into the chamber, upon which, the block and slide lock together ready for firing upon pressure on the trigger. After the last shot the empty magazine is ejected. Insertion of a full magazine closes the open slide automatically, loading the piece.

The pistol is marked on the right side of the slide "GRANT HAMMOND MFG. CORP., NEW HAVEN, CONN., USA". The top of the slide is marked "HAMMOND" in engraved letters. The left side is marked "PATENTED MAY 4, 1915, OTHER PATENTS PENDING". Number 8 is marked on the front of the frame, under the barrel.

A small number of these pistols, (reputed to be less than a dozen), were made up and submitted to the Navy Department for tests in 1917. The arm gave excellent results in functioning and accuracy. However any points of superiority that may have developed were not of sufficient relative importance to warrant the adoption of a new weapon during the war, or to render obsolete the large stocks of the excellent and reliable Model 1911 pistols.

SAVAGE Automatic Pistol, MODEL 1905.
Illustrated—Fig. 4, Plate 29.

Caliber .45, recoil operated, magazine fed, autoloading. The magazine has capacity for 8 rounds of rimless .45 caliber automatic cartridges. The barrel is five-and-one-quarter inches long and is rifled with six grooves. The total length of the pistol is nine inches. Weight with empty magazine is 2 pounds, 3½ ounces. Steel blade front sight; a U-notch rear sight is mounted on the slide. Blued finish, case-hardened hammer, milled at the top. Heavy milled corrugations on the slide give a firm grip and facilitate the withdrawal of the slide to the rear for initial loading. The pistol is equipped with a grip safety and a safety latch on the left side. The firing pin is pivoted on the hammer.

The pistol functions through the utilization of the recoil power of the cartridge to unlock, extract, eject and reload. When the cartridge is fired the slide moves backwards, turning the barrel partially to the right to unlock, extracts and ejects the empty shell. On its return to the forward position, the slide strips a live cartridge from the magazine, and drives it into the chamber, ready for firing upon pressure on the trigger. After firing the last shot, the slide remains open, ready for closing on the insertion of a full magazine.

The pistol illustrated is marked on the barrel "MANUFACTURED BY SAVAGE ARMS CO. IN UTICA, N. Y., U.S.A., NOV. 21, 1905. CAL. 45". Number 170 is stamped on the bottom of the slide under the hammer.

About two hundred of these pistols were made by the Savage Arms Company, but the arms not having been adopted by the government, the pistols never went into general production.

ILLUSTRATIVE
PLATES

PLATE 1

1. NORTH & CHENEY, Flintlock Pistol Model 1799.
2. HARPERS FERRY Flintlock Pistol Model 1806.
3. S. NORTH (BERLIN) Flintlock Pistol Model 1808, Navy (Right Side).
4. S. NORTH (BERLIN) Flintlock Pistol Model 1808, Navy (Left Side, showing Belt-hook).

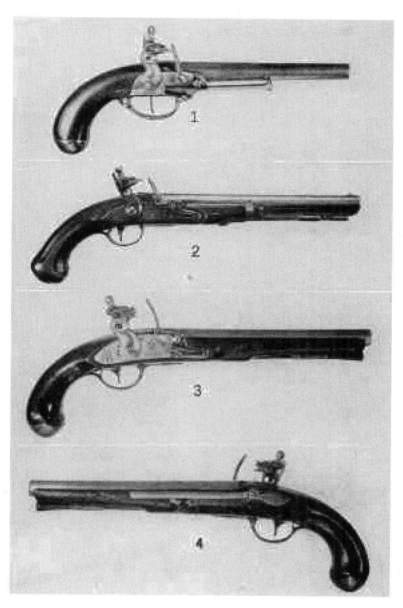

PLATE 1

PLATE 2

1. S. NORTH (BERLIN) FLINTLOCK PISTOL MODEL 1810.

2. S. NORTH (BERLIN) FLINTLOCK PISTOL MODEL 1810,
    (Transition Period Arm with Model 1813 Barrel Band.)

3. S. NORTH (BERLIN) FLINTLOCK PISTOL MODEL 1810,
    (Transition Period Arm with Model 1813 Iron Mountings.)

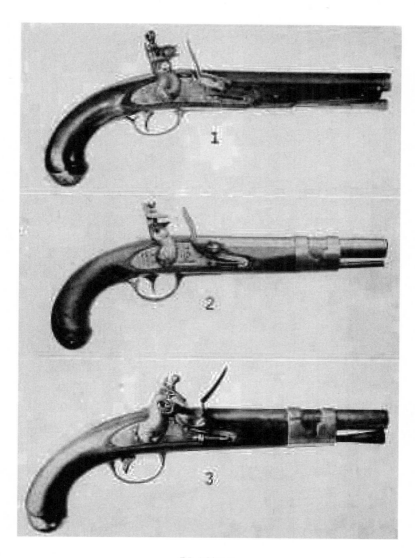

PLATE 2

PLATE 3

1. S. NORTH FLINTLOCK PISTOL MODEL 1813.
2. S. NORTH FLINTLOCK PISTOL MODEL 1813. (Marked "S. NORTH" and "US" in Two Horizontal Lines.)
3. S. NORTH FLINTLOCK PISTOL MODEL 1816.
4. SPRINGFIELD FLINTLOCK PISTOL MODEL 1818. (Goose-neck Hammer.)

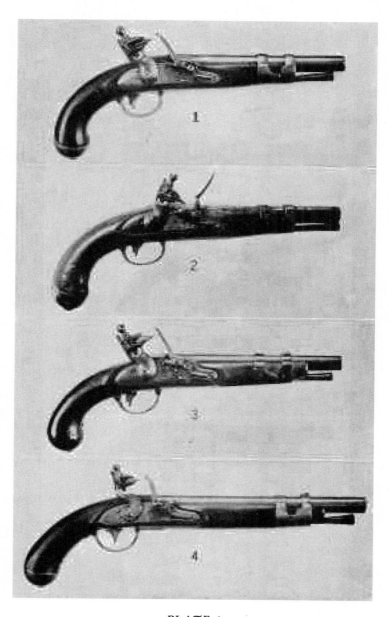

PLATE 3

PLATE 4

1. SPRINGFIELD FLINTLOCK PISTOL MODEL 1818, (Double-necked Hammer).

2. S. NORTH FLINTLOCK PISTOL MODEL 1819.

3. S. NORTH FLINTLOCK PISTOL MODEL 1826.

4. W. L. EVANS FLINTLOCK PISTOL MODEL 1826.

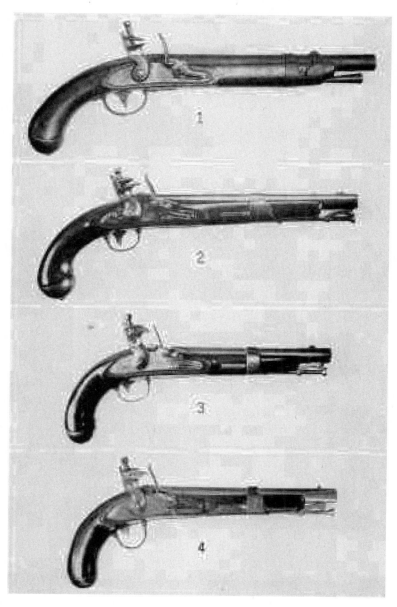

PLATE 4

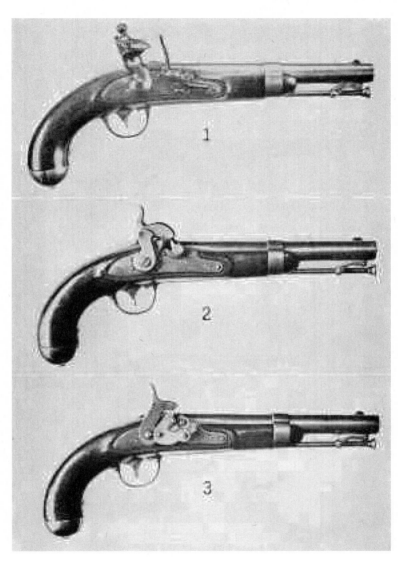

PLATE 5

PLATE 6

1. H. ASTON PERCUSSION PISTOL MODEL 1842.
2. AMES PERCUSSION PISTOL MODEL 1843, ARMY, NAVY.
3. SPRINGFIELD PISTOL-CARBINE MODEL 1855.

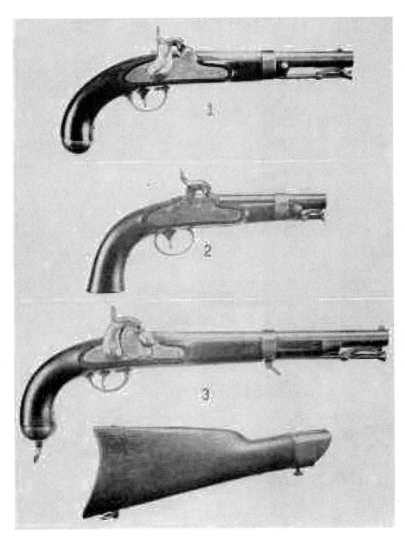

PLATE 6

PLATE 7

1. REMINGTON Single Shot Navy Pistol Model 1866.
2. REMINGTON Single Shot Navy Pistol Model 1867.
3. SPRINGFIELD Single Shot Army Pistol Model 1869.
4. REMINGTON Single Shot Army Pistol Model 1871.

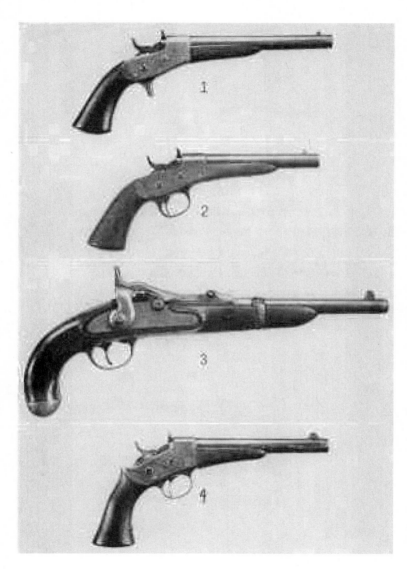

PLATE 7

PLATE 8

1. ANSTAT FLINTLOCK PISTOL.
2. BIELRY & CO. FLINTLOCK PISTOL.
3. C. BIRD & CO. FLINTLOCK PISTOL.
4. C. BIRD & CO. FLINTLOCK PISTOL., (Flat Butt).
5. BOOTH FLINTLOCK PISTOL.

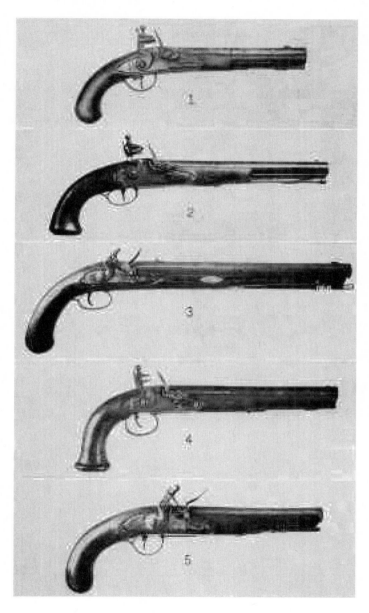

PLATE 8

PLATE 9

1. T. P. CHERINGTON Flintlock Pistol.
2. COUTTY Flintlock Pistol.
3. H. DERINGER Flintlock Pistol, Model 1808 Type.
4. JOHN DERR Percussion Pistol.
5. DREPPERD Percussion Pistol.

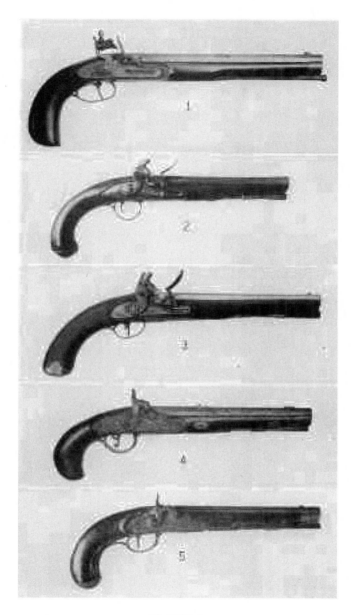

PLATE 9

PLATE 10

1. ELGIN Percussion Cutlass-Pistol, (C. B. ALLEN Make).

2. ELGIN Percussion Cutlass-Pistol, (MERRILL, MOS-MAN & BLAIR MAKE).

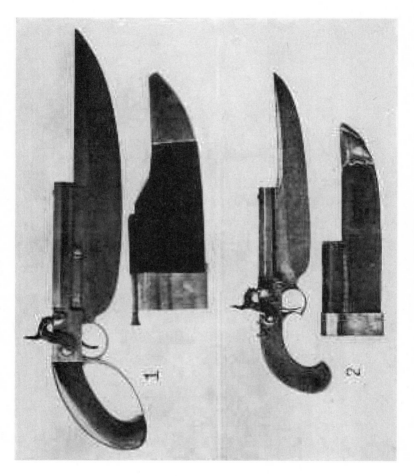

PLATE 10

PLATE 11

1. EVANS Flintlock Pistol, French Model 1805 Type.
2. T. FRENCH Flintlock Pistol, Model 1808 Type.
3. T. GRUBB Flintlock Pistol.
4. I. GUEST Flintlock Pistol, Model 1808 Type, (DREP-PERT Lock).

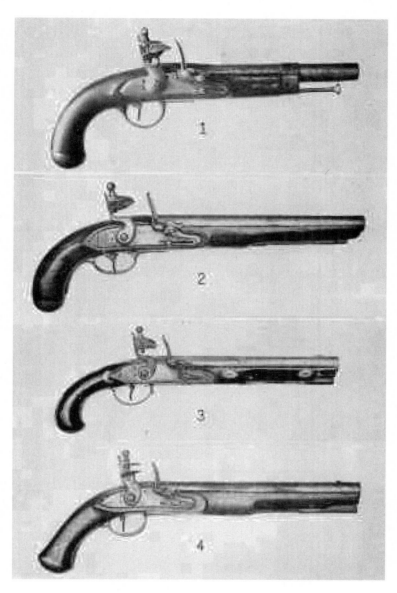

PLATE 11

PLATE 12

1. HALL Breech-loading Flintlock Pistol, Bronze Barrel and Breech. (Breech Closed).

2. HALL Breech-loading Flintlock Pistol, Bronze Barrel and Breech. (Breech Open).

3. HALL Breech-loading Flintlock Pistol, Iron Barrel and Breech.

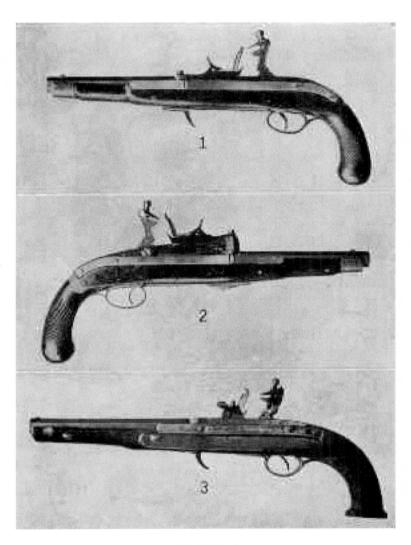

PLATE 12

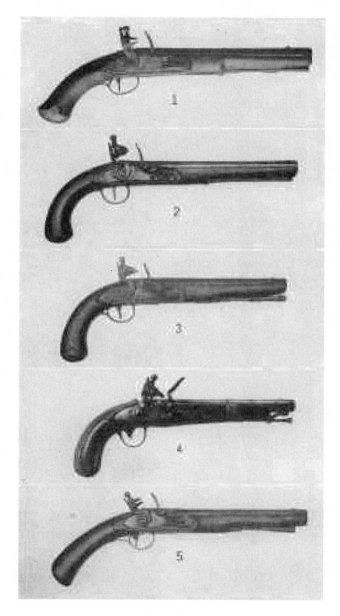

PLATE 13

PLATE 14

1. JACOB KUNTZ FLINTLOCK PISTOL.
2. LINDSAY PERCUSSION PISTOL.
3. MARSTON PERCUSSION PISTOL.
4. McK BROTHERS PERCUSSION PISTOL.
5. MILES FLINTLOCK PISTOL, MODEL 1808 TYPE.

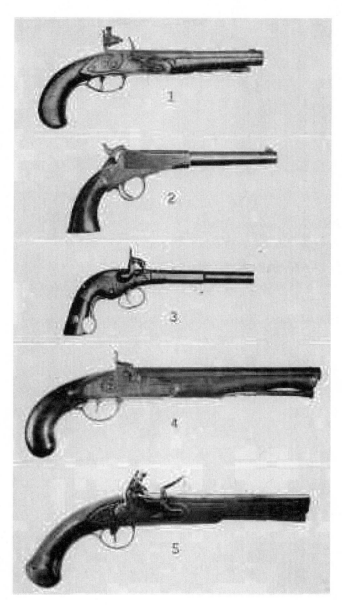

PLATE 14

PLATE 15

1. MILES FLINTLOCK PISTOL.

2. P. & D. MOLL FLINTLOCK PISTOL.

3. I. PERKINS FLINTLOCK PISTOL.

4. PERRY BREECH-LOADING PERCUSSION PISTOL, (With Capping Device).

5. PERRY BREECH-LOADING PERCUSSION PISTOL, (Without Capping Device).

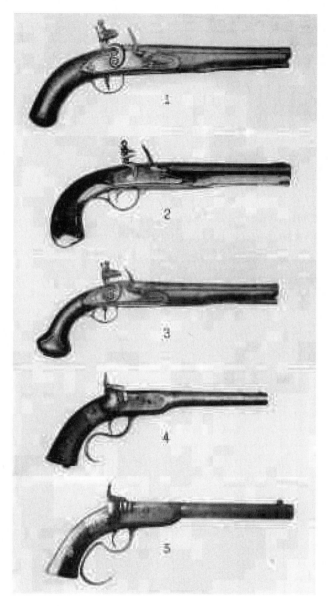

PLATE 15

PLATE 16

1. RICHMOND - VIRGINIA FLINTLOCK PISTOL.
2. SHARPS BREECH-LOADING PERCUSSION PISTOL.
3. VIRGINIA MANUFACTORY FLINTLOCK PISTOL.
4. J. WALSH FLINTLOCK PISTOL.
5. A. H. WATERS & CO. PERCUSSION PISTOL.

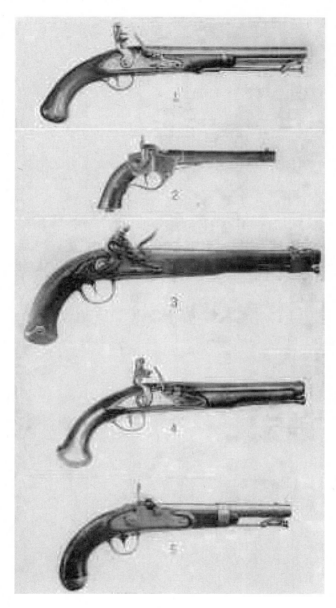

PLATE 16

PLATE 17

1. ADAMS Navy Percussion Revolver.
2. ALLEN & WHEELOCK Army Percussion Revolver.
3. ALLEN & WHEELOCK Navy Percussion Revolver, Side-Hammer Model.
4. ALSOP Navy Percussion Revolver.
5. BEALS Navy Percussion Revolver.
6. BUTTERFIELD Army Percussion Revolver.

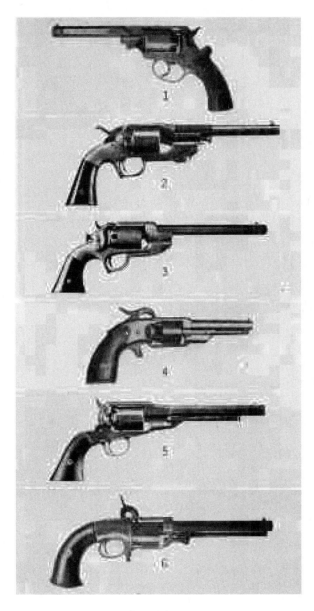

PLATE 17

PLATE 18

1. COLT ARMY REVOLVER MODEL 1847 (WHITNEYVILLE Make).

2. COLT ARMY PERCUSSION REVOLVER MODEL 1848 (DRA-GOON).

3. COLT ARMY PERCUSSION REVOLVER MODEL 1848 (DRA-GOON) WITH CANTEEN PISTOL-CARBINE STOCK.

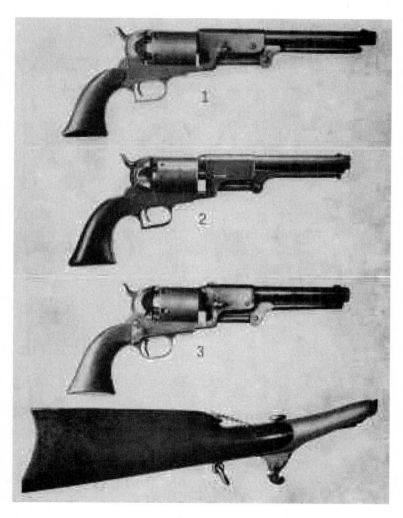

PLATE 18

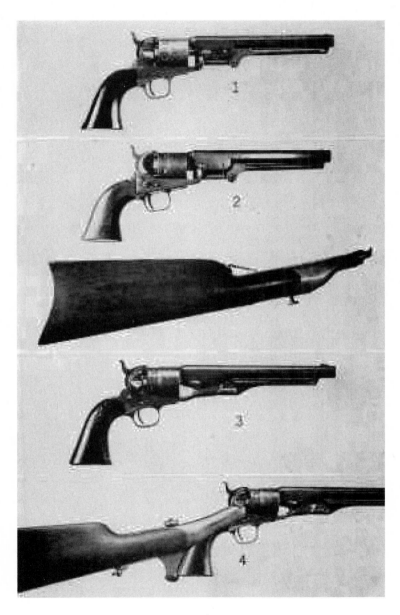

PLATE 19

PLATE 20

1. COLT Army Percussion Revolver Model 1860, Full Fluted Cylinder.

2. COLT Army Revolver Model 1860, Conversion to Cartridge.

3. COLT Navy Percussion Revolver Model 1861.

4. COOPER Navy Percussion Revolver.

5. FREEMAN Army Percussion Revolver.

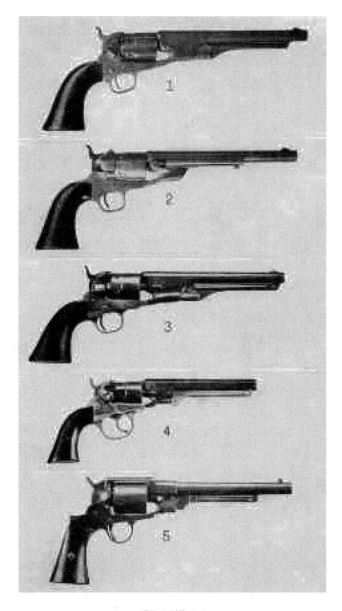

PLATE 20

PLATE 21

1. JOSLYN ARMY PERCUSSION REVOLVER.
2. MANHATTAN NAVY PERCUSSION REVOLVER.
3. METROPOLITAN NAVY PERCUSSION REVOLVER.
4. METROPOLITAN NAVY PERCUSSION REVOLVER (Pocket Navy).
5. PETTINGILL ARMY PERCUSSION REVOLVER.
6. PETTINGILL NAVY PERCUSSION REVOLVER.

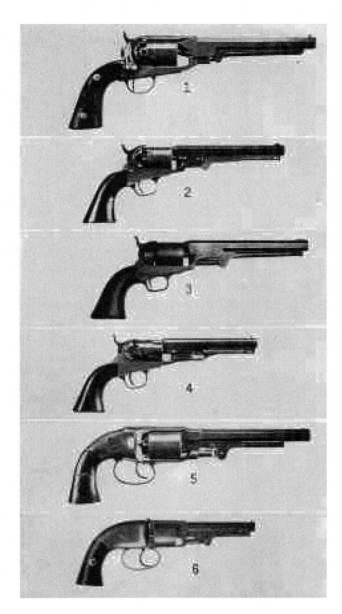

PLATE 21

PLATE 22

1. REMINGTON Army Percussion Revolver, Model 1861.
2. REMINGTON Army Percussion Revolver, New Model.
3. & 4. REMINGTON NEW MODEL Army Percussion Revolvers Converted to Cartridge.
5. REMINGTON-RIDER Navy Percussion Revolver.
6. ROGERS & SPENCER Army Percussion Revolver.

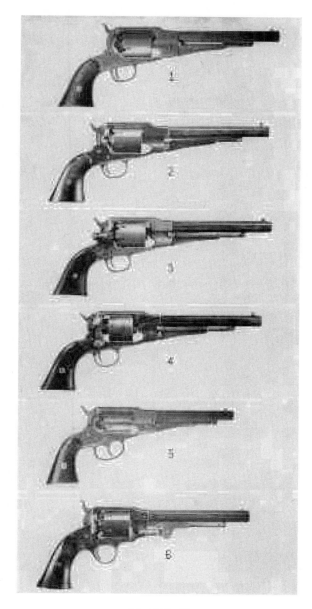

PLATE 22

PLATE 23

1. SAVAGE-NORTH Navy Percussion Revolver.
2. SAVAGE Navy Percussion Revolver.
3. STARR Army Percussion Revolver—Single Action.
4. STARR Army Percussion Revolver—Double Action.
5. STARR Navy Percussion Revolver—Double Action.

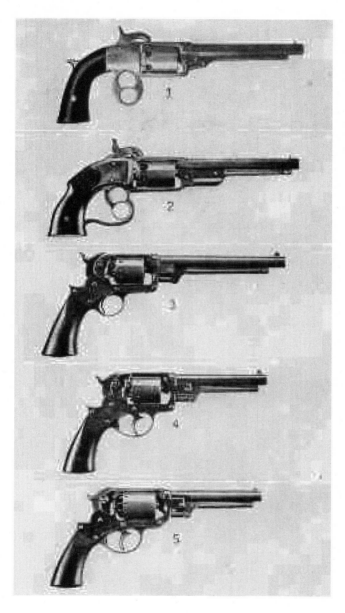

PLATE 23

PLATE 24

1. WALCH 12-SHOT PERCUSSION NAVY REVOLVER. (Spur Grip)
2. WALCH 12-SHOT PERCUSSION NAVY REVOLVER.
3. WARNER NAVY PERCUSSION REVOLVER.
4. WESSON & LEAVITT ARMY PERCUSSION REVOLVER.
5. WESSON & LEAVITT PERCUSSION REVOLVER.
6. WHITNEY NAVY PERCUSSION REVOLVER.

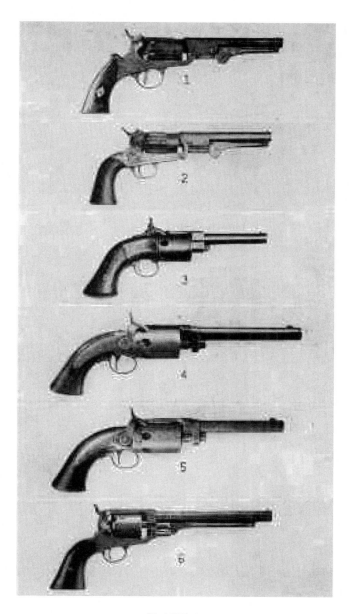

PLATE 24

PLATE 25

1. BACON Navy Model Revolver.

2. COLT Army Revolver Model 1872.

3. COLT Army Revolver Model 1878.

4. COLT Army Revolver Model 1878 (With Alaska Trigger Guard.)

5. COLT Army Revolver Model 1896.

6. COLT Army Revolver Model 1909.

PLATE 25

PLATE 26

1. COLT Army Revolver Model 1917.
2. FOREHAND & WADSWORTH Army Revolver.
3. HOPKINS & ALLEN XL Navy Revolver.
4. MERWIN & HULBERT Army Revolver Model 1876.
5. MERWIN & HULBERT Army Pocket Revolver.

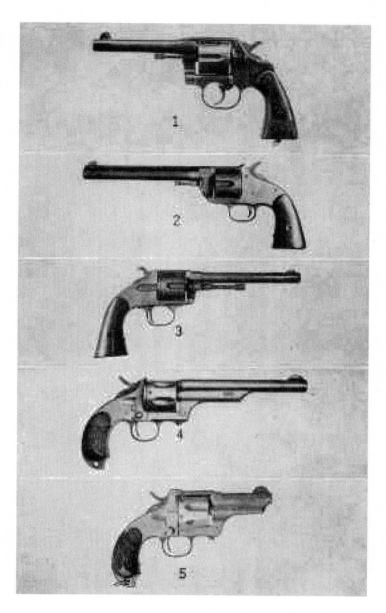

PLATE 26

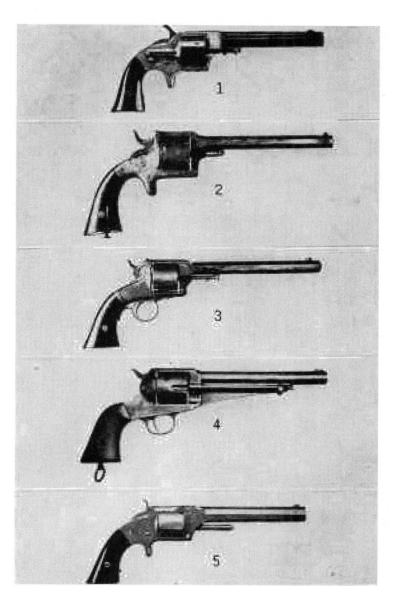

PLATE 27

PLATE 28

SMITH & WESSON REVOLVERS

1. S. & W. ARMY REVOLVER MODEL 1869, AMERICAN.

2. S. & W. ARMY REVOLVER MODEL 1875 SCHOFIELD, EARLY LATCH.

3. S. & W. ARMY REVOLVER MODEL 1875 SCHOFIELD, LATER LATCH.

4. S. & W. NAVY REVOLVER MODEL 1881, (NEW MODEL NAVY NO. 3).

5. S. & W. ARMY, NAVY REVOLVER MODEL 1899.

6. S. & W. ARMY REVOLVER MODEL 1917.

PLATE 28

PLATE 29

1. COLT Automatic Pistol Military Model 1902.

2. COLT Automatic Pistol Model 1911-A1.

3. GRANT-HAMMOND Automatic Pistol, (Navy Experimental).

4. SAVAGE Automatic Pistol Model 1905.

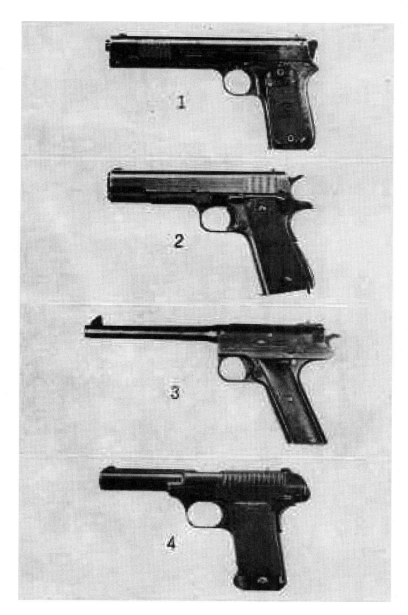

PLATE 29